Springer Aerospace Technology

The series explores the technology and the science related to the aircraft and spacecraft including concept, design, assembly, control and maintenance. The topics cover aircraft, missiles, space vehicles, aircraft engines and propulsion units. The volumes of the series present the fundamentals, the applications and the advances in all the fields related to aerospace engineering, including:

- structural analysis,
- aerodynamics,
- aeroelasticity,
- aeroacoustics,
- flight mechanics and dynamics
- orbital maneuvers,
- avionics,
- systems design,
- materials technology,
- launch technology,
- payload and satellite technology,
- space industry, medicine and biology.

The series' scope includes monographs, professional books, advanced textbooks, as well as selected contributions from specialized conferences and workshops.

The volumes of the series are single-blind peer-reviewed.

To submit a proposal or request further information, please contact: Mr. Pierpaolo Riva at pierpaolo.riva@springer.com (Europe and Americas) Mr. Mengchu Huang at mengchu.huang@springer.com (China)

The series is indexed in Scopus and Compendex

Alexander S. Sigov

Multilayer Magnetic Nanostructures

Properties and Applications

Springer

Alexander S. Sigov
Department of Nanoelectronics
MIREA—Russian Technological
University
Moscow, Russia

Russian Academy of Sciences
Moscow, Russia

ISSN 1869-1730 ISSN 1869-1749 (electronic)
Springer Aerospace Technology
ISBN 978-981-19-6245-5 ISBN 978-981-19-6246-2 (eBook)
https://doi.org/10.1007/978-981-19-6246-2

This Springer imprint is published by the registered company Springer Nature Singapore Pte Ltd.
The registered company address is: 152 Beach Road, #21-01/04 Gateway East, Singapore 189721, Singapore

The book is dedicated to my dear friend and colleague Alexander Igorevich Morosov, Professor, Doctor of Physical and Mathematical Sciences, an outstanding Russian scientist, who passed away on October 12, 2020.

Introduction

A surge of interest in multilayer magnetic structures, whose layer thicknesses range from tenths to tens of nanometers, is associated with the discovery of phenomena of giant magnetoresistance (GMR) in 1988. It exceptionally quickly found its practical application in the reading heads of hard drives, which already allowed to increase their capacity from hundreds of megabytes to tens in the early 90s of the last century, and then hundreds of gigabytes, and thus significantly expand the capabilities of computer technology used, including in aviation computing and navigation systems used both onboard aircraft and in ground-based flight support.

The broad technical applications of this phenomenon attracted the attention of the scientific community, which was reflected in the award of the Nobel Prize in Physics in 2007 to the heads of scientific groups who discovered the phenomenon of GMR—A. Fert and P. Grunberg.

In the field of basic research, the discovery of GMR stimulated a broad study of systems containing nanolayers of magnetic materials, which led to the discovery in 1995 of the phenomena of tunneling magnetoresistance (TMR) spin-polarized current magnetization reversal, and several others. The direction of electronics, called "spintronics", was formed, which combined the phenomena and devices using spin-polarized currents, that is, the transfer of information using spin, while traditional electronics uses the phenomenon of electric charge transfer.

In the applied field, the discovery of the phenomena of GMR and TMR stimulated the creation of magnetoresistive memory (MRAM), which is energy-independent, has a speed of about nanoseconds, and in the future can replace both hard disks and flash and a semiconductor memory. Prototypes of such memory have been developed, and their serial production is being mastered. The main disadvantages of MRAM are the complexity of integration with traditional semiconductor electronics technologies and the high cost.

Contents

About the Author

Alexander S. Sigov Academician of the Russian Academy of Sciences, Honored Scientist of the Russian Federation, Laureate of the State of the Russian Federation Prize in the field of science and technology, Laureate of the Government of the Russian Federation Prize in the field of science and technology, Laureate of the UNESCO Medal for Contribution to the Development of Nanoscience and Nanotechnology, Professor, Doctor of Physical and Mathematical Sciences.

Abbreviations

AMR	Anisotropic magnetoresistance
EC	Electronic computer
GMR	Giant magnetoresistance
MRAM	Magnetoresistive memory
PSV	Pseudo-spin valve
RKKY	Ruderman–Kittel–Kasuya–Yoshida interaction
SHF	Superhigh frequency
SPLEEM	Spin-polarized low-energy electron microscopy
SPRAM	Spin-transfer torque random access memory
SRAM	Static random access memory
TMR	Tunneling magnetoresistance

Chapter 1
Physical Foundations for the Formation of Magnetic Nanostructures

Check for updates

1.1 The Phenomenon of Giant Magnetoresistance

What is the reason for such a breakthrough with the density of information recording on hard disks? The information on the hard disk is recorded at (ferro-) magnetic tracks separated by non-magnetic gaps (Fig. 1.1). The bit of information is recorded at the track as a domain. It is just the direction of magnetization in the domain that carries the information: "zero", for example, in the case of a longitudinal record (the magnetization of the domains is parallel to the surface of the track) corresponds to the magnetization directed to the right, and "one"—to the left. The domain size is limited from below (except for the superparamagnetic limit, which will be discussed further) by the condition that the strength of the magnetic field created by the domain (Fig. 1.1) has a sufficient magnitude to register it with the reading head of the hard disk. In other words, the size of the domain is determined by the sensitivity of the read head.

In the late 80 s–early 90 s of the last century, hard drives used reading heads based on the effect of the anisotropic magnetoresistance (AMR), which consists of the fact that a ferromagnet's resistance depends on the angle φ between its magnetization and the electric current direction $\rho = \rho(\varphi)$. The magnitude of the effect $A = \frac{\rho_{\max} - \rho_{\min}}{\rho_{\max}} = \frac{\Delta\rho}{\rho_{\max}}$ was tenths of a percent.

The AMR-based reading head used four pieces of ferromagnet with a slight misorientation of the easy magnetization axes. The head sliding over the disk's surface changes its resistance under the influence of the domain field at the magnetic track. This change depends on the direction of the field strength, which makes it possible to distinguish between the written "0" and "1". The small value of parameter A and the presence of several ferromagnetic pieces in the head impose a lower limit on the minimum size of the domain at the magnetic track.

Let us consider the essence of the GMS effect based on a spin-valve system consisting of two ferromagnetic metal layers separated by a metal non-ferromagnetic layer (non-magnetically ordered—*Co-Cu-Co* or antiferromagnetic—*Fe-Cr-Fe*) (Fig. 1.2). The thicknesses of the layers should be from fractions of a

A. S. Sigov, *Multilayer Magnetic Nanostructures*, Springer Aerospace Technology,
https://doi.org/10.1007/978-981-19-6246-2_1

Fig. 1.1 The domain at the hard disk track

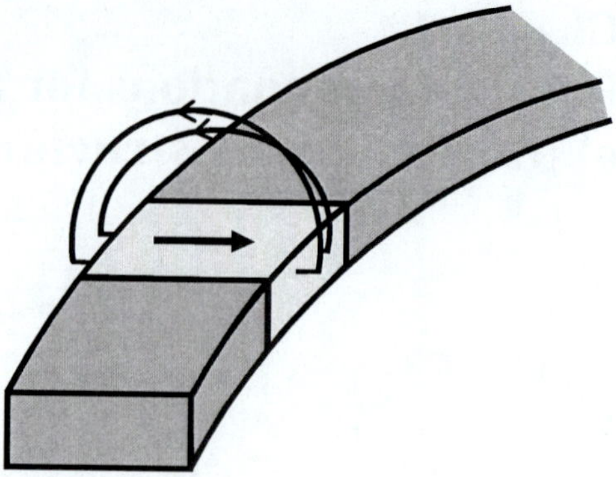

nanometer up to units of nanometers, that is, the spin gate is a magnetic nanostructure. The current can flow through the spin valve parallel to the layers (current in plane—CIP-orientation) or perpendicular to them (current perpendicular to plane—CPP-orientation). As will be shown further, the quality of the interfaces between the layers plays a fundamental role.

In a non-magnetic layer, the exchange interaction between the layers is due to an indirect exchange through conduction electrons, i.e., the Ruderman-Kittel-Kasuya-Yoshida (RKKY) interaction. The exchange integral in the case of the RKKY interaction oscillates with the thickness of the interlayer d and decreases as d^{-2} (Fig. 1.3):

Fig. 1.2 Spin-valve systems *Fe-Cr-Fe* (**a**) and *Co-Cu-Co* (**b**)

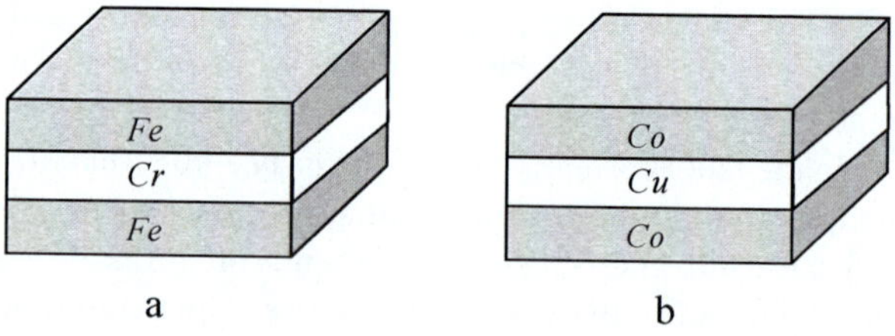

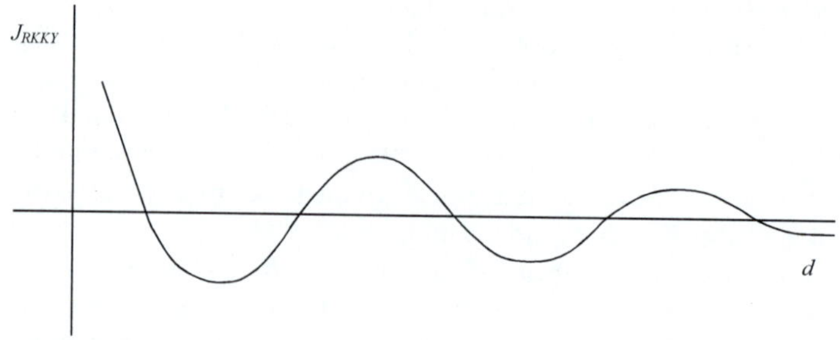

Fig. 1.3 The dependence of the interlayer interaction on the thickness of the interface

$$J_{\mathrm{RKKY}} = \frac{C \sin(2k_F \mathrm{d})}{d^2}, \tag{1.1}$$

where κ_F is the Fermi wave vector of electrons in the interlayer. The energy of the exchange interaction of uniformly magnetized ferromagnetic layers with magnetizations $\vec{M}_1$ and $\vec{M}_2$ has the form:

$$E_{\mathrm{ex}} = -J_{\mathrm{RKKY}}\left(\vec{M}_1, \vec{M}_2\right) \tag{1.2}$$

If $J_{\mathrm{RKKY}} > 0$, then the parallel orientation of the magnetizations is advantageous, and if $J_{\mathrm{RKKY}} < 0$, then the antiparallel direction is beneficial.

You can choose such a thickness of the layer d that the value J_{RKKY} is negative. Then, in the absence of a magnetic field, the magnetizations of the ferromagnetic layers are oriented antiparallel.

When the magnetic field is turned on, each magnetization tends to orient itself along the field, but the exchange interaction between them prevents this. As a result, they will be oriented as shown in Fig. 1.4. As the magnetic field increases, the angle between the magnetizations decreases. This behavior is similar to the conduct of the magnetizations of the antiferromagnet sublattices in the spin-flop phase. That will continue until the sum of the energies of the two ferromagnetic layers in the external field:

$$E_B^i = -V_i\left(\vec{B}_0, \vec{M}_{\mathrm{i}}\right), \tag{1.3}$$

where V_i is the volume i of the ferromagnetic layer ($i = 1, 2$), and $\vec{E}_0$ is the induction of the external magnetic field and will not exceed in the field E_{sat} the energy of the exchange interaction between the layers in the area. In the fields $B_0 > B_{\mathrm{sat}}$, the magnetizations of the layers will be oriented parallel to each other, and the total magnetization will reach saturation.

The dependence of the resistance of the three-layer structure on the value E_0 is shown in Fig. 1.5. It is easy to see that in the case of the antiparallel orientation of the magnetizations, the resistance of the spin-valve structure R_{AP} is higher than in the case of parallel one (R_P). The value $A = \frac{R_{\mathrm{AP}} - R_P}{R_{\mathrm{AP}}}$ is of the order of 10%, that is, two orders of magnitude higher than the corresponding value in the case of AMR.

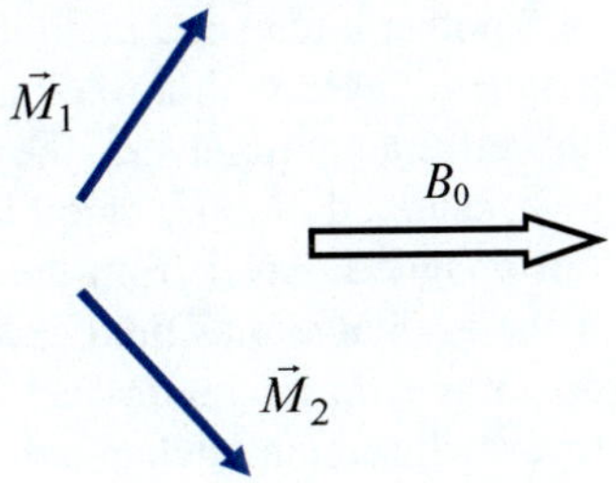

Fig. 1.4 Spin-flop orientation of the magnetizations of ferromagnetic layers

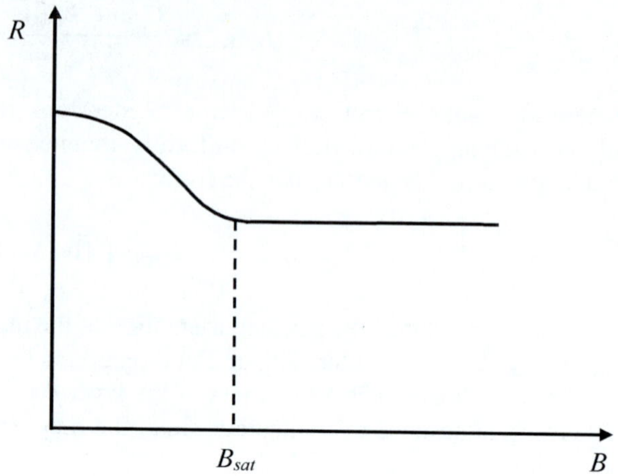

Fig. 1.5 The dependence of the resistance of the spin-valve structure on the magnitude of the magnetic field

It should be noted that GMR is observed not only in spin-valve structures but also in multilayer structures consisting of dozens of alternating ferromagnetic and non-ferromagnetic metal layers. The GMR effect was first discovered just in such multilayer structures.

Reading heads based on GMR were developed and put into production two years after discovering this effect when scientists have not yet had time to understand the nature of this phenomenon fully.

To create a reading head based on GMR, it is necessary to ensure that the field created by the domain at the magnetic track would reorient the magnetization of only one of the ferromagnetic layers of the spin-valve structure. Then, being above the bit in which "0" is written, the head will be in a state with higher resistance R_{AP}, and being above the bit in which "1" is written, it will be in a condition with less resistance state R_P, which makes it possible to distinguish them.

How can we fix the magnetization of one of the ferromagnetic layers of the spin-valve structure? That can be implemented in several ways. The simplest solution (from the point of view of physics, but not technology) is to make layers of materials with different coercive forces so that the domain field can switch the magnetization of one ferromagnetic layer, but is not sufficient to change the magnetization of the second. Sometimes, you can use a layer of the same material but with a larger thickness.

Another interesting method is using an antiferromagnetic layer adjacent to a ferromagnetic layer with a fixed magnetization. In the interaction of ferromagnetic and antiferromagnetic layers, the phenomenon of the exchange bias occurs, which will be considered in more detail later. It consists in the shift of the hysteresis loop of the ferromagnetic layer from the position symmetrical in the magnetic field (Fig. 1.6). If the exchange bias field exceeds the maximum values of the fields created by the domains at the magnetic track of the hard disk, the magnetization of the ferromagnetic layer will remain unchanged when reading both "0" and "1".

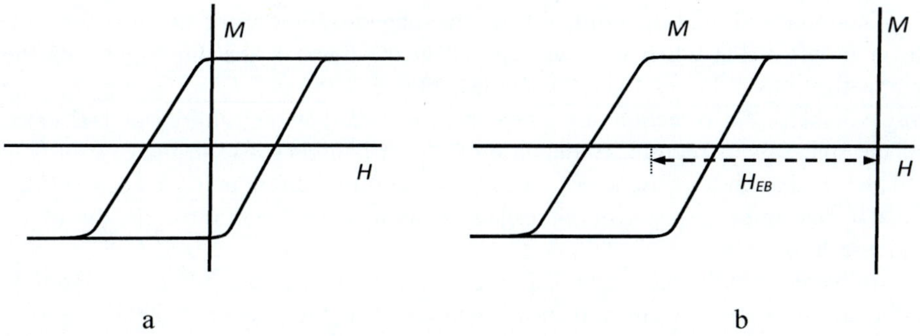

Fig. 1.6 The hysteresis loop in the absence (**a**) and with account (**b**) of the interaction of the ferromagnet with the antiferromagnetic layer, H_{EB} is the value of the exchange bias

A schematic representation of the hard disk reading head based on GMR is shown in Fig. 1.7. The principle of its operation is elementary. The magnetic domain over which it is currently located creates a magnetic field parallel to the plane of the disk, which acts on the spin-valve system of the reading head. This three-layer system consists of two ferromagnetic layers with different crystal anisotropies, separated by a non-magnetic interlayer. The magnetically rigid layer 1 is magnetized parallel (or antiparallel) to the field created by the domain, and its coercive field exceeds the domain field. As a result, its magnetization does not change under the influence of the domain field. The magnetically soft layer 2 has a smaller coercive field and is magnetized in a direction perpendicular to the magnetization of the rigid layer and the field generated by the domain. That is, the axes of easy magnetization of layers 1 and 2 are mutually perpendicular. This orientation corresponds to the highest sensitivity with respect to the external field (*the* d*R*/d*H derivative* is maximal).

Fig. 1.7 Hard disk read head based on the GMR effect

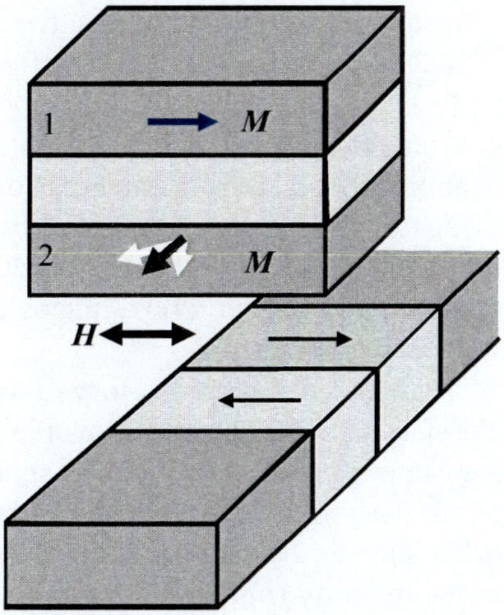

Under the action of the domain field, the magnetization of the magnetically soft layer turns so that when it is written "0" in the domain, the angle between the magnetizations of layers 1 and 2 becomes obtuse, and in the case of writing "1" the angle is sharp. The direction of the reversal is recorded by the value of the resistance of the spin-valve structure. In the case of "0", it is higher, and in the case of "1", it is lower than in the absence of a signal. A constant current is passed through the head to determine the value of the resistance, fixing the value of the derivative of the voltage drop.

An increase in the parameter A in the case of GMR by two orders of magnitude compared to AMR case allowed one to significantly reduce the width of the magnetic tracks and the size of the domains at them, which led to a significant (by 2–3 orders of magnitude) increase in the capacity of hard disks.

At the moment, to increase the recording density, a gradual transition is being made to hard drives based *on Co-Pt* and *Fe-Pt alloys* with perpendicular recording. These alloys have a large uniaxial crystallographic anisotropy, with an easy axis perpendicular to the track plane. In this case, the magnetization of the domain is perpendicular to the plane of the hard disk, and the recording of "0" and "1", as in the case of parallel recording, differs in the direction of the magnetization vector. The reading head for the case of perpendicular writing can be easily obtained from the considered one by simply turning it by 90°, so that the magnetization of the magnetically rigid layer is perpendicular to the plane of the hard disk.

1.2 GMR Theory

Let us consider first the explanation (not entirely correct, but simple) that roams all the tutorials and sites on the Internet. It proceeds from the correct assumption that in ferromagnetic materials, where the law of electron dispersion is split into subzones corresponding to electrons with opposite spin projections on the direction of magnetization, such values as free path times, the density of states on the Fermi surface, etc., for electrons with opposite spins can differ significantly.

Consequently, the contributions of these two types of electrons to the electrical conductivity of a ferromagnet may differ significantly. Let us denote the contribution to the resistivity from electrons whose magnetic moment is parallel to the magnetization of the ferromagnet, as ρ_M^{-1}, such electrons are the majority. And the contribution of minority electrons, whose magnetic moments are antiparallel to the magnetization, will be denoted as ρ_{Min}^{-1}.

The characteristic trajectory of an electron in a spin-valve structure in the quasi-classical approximation is shown in Fig. 1.8. In this case, in the first approximation, we can neglect the processes of spin-flip and assume that when an electron is scattered, the projection of its spin remains unchanged. Then, in the case of parallel orientation of the magnetizations of the ferromagnetic layers, the electron belonging to the majority (minority) in the first ferromagnetic layer will not change its "fractional" membership, falling into the second ferromagnetic layer: the electron of

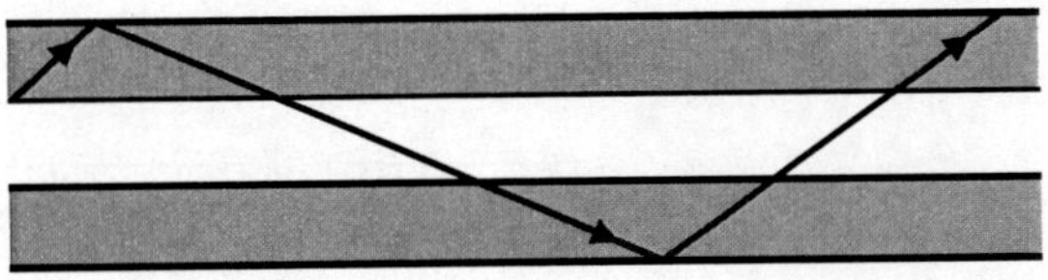

Fig. 1.8 The trajectory of an electron in a spin-valve structure

the "majority" will remain the electron of the "majority", and the electron of the "minority" will remain the electron of the "minority".

Since the electron passes through the layers sequentially on its way, for the "majority" electron, the resistance of the entire spin-valve structure $R^{\text{sp - v}}$ is

$$R_M^{\text{sp - v}} a(\rho_M + \rho_M) = 2a\rho_M, \tag{1.4}$$

where a is a geometric factor that considers the thicknesses of the ferromagnetic layers, which we consider to be the same R^{sp-v}. For simplicity, we neglect the contribution of the interlayer to the resistance [1].

Similarly, for minority electrons, we obtain:

$$R_{\text{Min}}^{\text{sp - v}} = 2a\rho_{\text{Min}}. \tag{1.5}$$

Since the contributions of two types of electrons to the electrical conductivity are additive, then

$$\frac{1}{R_p} = \frac{1}{2a\rho_{\text{Min}}} + \frac{1}{2a\rho_M} = \frac{\rho_M + \rho_{\text{Min}}}{2a\rho_M\rho_{\text{Min}}}. \tag{1.6}$$

Suppose the magnetizations of the ferromagnetic layers are antiparallel. In that case, the electron that belonged to the majority in the first ferromagnetic layer, falling into the second ferromagnetic layer, immediately passes into the "opposition", that is, it will belong to the minority. With the "minority" electron, the opposite happens—it joins to the "majority".

Thus, for electrons of any sort (the first layer gives the sort name):

$$R_{M(\text{Min})}^{\text{sp - v}} = a(\rho_M + \rho_{\text{Min}}). \tag{1.7}$$

The total electrical conductivity is twice as high as the electrical conductivity produced by one sort of electrons, and

$$R_{\text{AP}} = \frac{a}{2}(\rho_M + \rho_{\text{Min}}). \tag{1.8}$$

The value is

$$\Delta R = R_{\text{AP}} - R_P = \frac{a(\rho_M - \rho_{\text{Min}})^2}{2(\rho_M + \rho_{\text{Min}})} \tag{1.9}$$

and

$$A = \frac{\Delta R}{R_{\mathrm{AP}}} = \left(\frac{\rho_M - \rho_{\mathrm{Min}}}{\rho_M + \rho_{\mathrm{Min}}} \right)^2. \tag{1.10}$$

This simple proof of the negative sign of the giant magnetoresistance (in the field, the resistance is less than without it) suffers from one significant drawback. The concept of local electrical conductivity can be introduced when the characteristic size of the problem (in our case, the thickness of the layers) is much greater than the free path of the charge carrier. Otherwise, Ohm's law takes a non-local form. The free path of electrons at room temperature is tens of nanometers. With layer thicknesses of units (fractions) of nanometers, it far exceeds the layer thicknesses. Therefore, the above simple explanation, unfortunately, does not apply even qualitatively.

The correct explanation is much more complicated. Let us pay attention to the potential relief in which the electron of the majority (minority) of the first layer is moving in the cases of parallel and antiparallel orientation of the magnetizations of the ferromagnetic layers (Fig. 1.9). It is easy to see that the potential reliefs for these two cases are different for each sort of electrons. Therefore, solving the Schrödinger equation, we obtain different Ψ-functions of the "majority" and "minority" electrons in parallel and antiparallel orientation of the magnetizations. Since the functions Ψ are different, we obtain different densities of states of electrons and matrix elements which determine the probability of scattering of charge carriers from phonons and impurities, and, consequently, different free path times and the resistance values of the spin-valve structure. Obtaining a value ΔR requires relatively complex calculations, and the sign of this value cannot be found by simple reasoning [2].

It should be noted that the actual boundaries of the layers are rough, which makes an additional contribution to the resistance of multilayer magnetic structures.

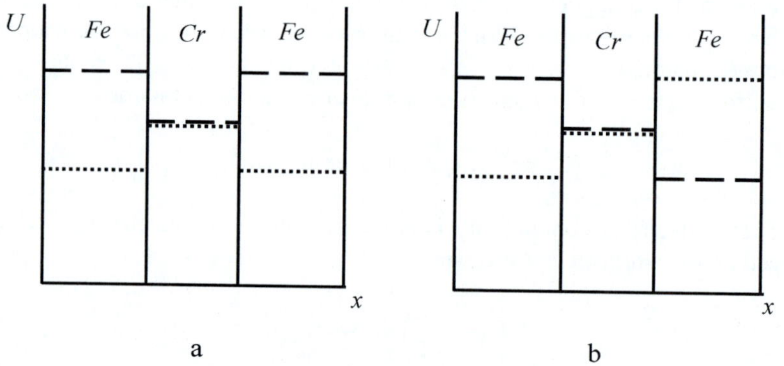

Fig. 1.9 The potential energy relief for the majority (points) and minority (dashes) electrons in the case of parallel (**a**) and antiparallel (**b**) orientation of the magnetizations in the spin-valve structure

1.3 Tunnel Magnetoresistance

So far, we have considered multilayer structures containing only metal layers. And what happens if the metal layer is replaced with a dielectric one? In CIP geometry, the current flows along the layers will cause the current to flow independently through the ferromagnetic layers, and the magnetoresistance effect will almost disappear.

If the current is applied perpendicular to the layers (CPR-geometry), then the dielectric layer will represent a potential barrier for the charge carriers. The tunneling of electrons will cause the flow of current through this barrier. For the tunnel current to be noticeable, the thickness of the interlayer must be $\leq$1 nm.

In the case of a magnetic tunnel structure consisting of two ferromagnetic metal layers separated by a dielectric layer, the quality of the interface plays an even more critical role than in the case of a spin-valve structure. Attempts to obtain high-quality and reproducible forms using aluminum oxide Al_2O_3 as a dielectric have not been successful. Only after replacing it with MgO (magnesium oxide) was it possible to solve the problem.

As in the case of GMR, it turned out that the resistance of the tunnel magnetic structure at parallel magnetizations of ferromagnetic layers is lower than at antiparallel ones. The value of the tunnel magnetoresistance (TMR) is essential $A = \frac{\Delta R}{R_{\mathrm{AP}}} = \frac{R_{\mathrm{AP}}-R_P}{R_{\mathrm{AP}}} \sim 70\%$.

Let us consider the causes of this phenomenon qualitatively. The electron tunnels from the cathode to the anode. The probability of elastic tunneling of the charge carrier without changing the spin projection, and, consequently, the main contribution to the tunneling current, is proportional to the factor

$$\nu_\sigma(\varepsilon)F_0(\varepsilon)(1 - F_0(\varepsilon + \mathrm{e}V))\nu_\sigma(\varepsilon + \mathrm{e}V), \quad (1.11)$$

where $F_0(\varepsilon)$ is the Fermi–Dirac distribution function, $\upsilon_\sigma(\varepsilon)$ is the density of the majority (minority) electron states ($\sigma = M$, Min) in a metal ferromagnet, and e is the elementary charge. If V the potential difference is applied to the structure (it is important to note that all of it falls on the dielectric layer), then the electron that had power in the cathode ε enters the anode with power $\varepsilon + \mathrm{e}V$. According to the Pauli principle, tunneling is possible only in the presence of a free state with such power.

This factor should be integrated over all electron energies ε and summed over σ.

It is easy to see that the factor $F_0(\varepsilon)(1 - F_0(\varepsilon + \mathrm{e}V))$ is significantly different from zero in the range $\max(T, \mathrm{e}V)$ near the Fermi power ε_F (T—temperature in power units). Assuming that $\varepsilon_F >> \max(T, \mathrm{e}V)$, we obtain that in the case of parallel magnetizations of ferromagnetic layers

$$I_P \infty \nu_M^2(\varepsilon_F) + \nu_{\mathrm{Min}}^2(\varepsilon_F), \quad (1.12)$$

Here, we assume that both ferromagnetic layers are made of the same material.

In the case of antiparallel orientation of the magnetizations, we have [1]

$$I_{AP} \propto 2\nu_M(\varepsilon_F)\nu_{Min}(\varepsilon_F). \tag{1.13}$$

It is easy to see that $I_P \geq I_{AP}$, and, therefore, $R_P < R_{AP}$.

1.4 Spin-Polarized Current

As already noted, in a ferromagnet, the concentration and density of states on the Fermi surface for electrons with opposite spin projections on the direction of magnetization of the ferromagnet are different. Therefore, when an electric current is created in a ferromagnet, say, by an electric field, the current will be spin-polarized; that is, there will be more charge carriers with one direction of spin than with the other. As a result, such a spin-polarized current and the charge will carry the spin—proper moment of the amount of motion. Spin transfer can be used to transfer information. The newly created science of spintronics is exploring these possibilities.

Let us now consider the case when a spin-polarized current, leaving a thick ferromagnetic layer, passes through a layer of non-magnetic metal and a second thin layer of a ferromagnet (Fig. 1.10), whose thicknesses are much smaller than the characteristic length l_{sf} at which the spin relaxation occurs. When $I < 0$ spin-polarized electrons, leaving a thick ferromagnetic layer, fall into a thin ferromagnetic layer, the spin-polarized current, along with the spin, carries the associated magnetic moment.

If the magnetization of a thin ferromagnetic layer is non-collinear to the magnetization of a thick layer, then an effective moment (spin-transfer torque effect) will act on it from the polarized electrons entering the layer. Consider the reasons for its occurrence. The states of the electrons entering the thin ferromagnetic layer are characterized by the spin projection $\pm\frac{1}{2}$ on magnetization of the thick ferromagnetic layer. They are not eigenstates in the thin ferromagnetic layer (unless the magnetization of the thick layer forms an angle of 0° or 180° with the magnetization of the thin layer). Therefore, under the action of the exchange interaction between the spins of the incoming electrons and the spins of the electrons responsible for the magnetization of the thin layer, at several interatomic distances, these states relax to the eigenstates of the Hamiltonian of the thin ferromagnetic layer with a new spin quantization axis. In this case, the total spin of the incoming electrons changes. As is known from mechanics, the difference in the moment of the amount of motion is

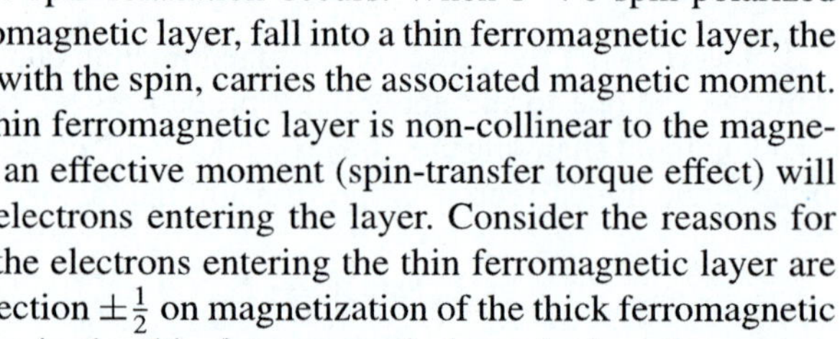

Fig. 1.10 Layer configuration: *F*—ferromagnetic, *M*—non-magnetic metal layers

associated with the action of the moment of forces on the object. But if the moment of forces acts on the incoming electrons from the side of the thin ferromagnetic layer, then, according to Newton's third law, the opposite moment of forces will work on the magnetization of the thin ferromagnetic layer from the side of the incoming electrons.

Phenomenologically, the magnetization behavior of a thin ferromagnetic layer can be described on the basis of the Landau-Lifshitz-Hilbert equation, which is used to consider the dynamics of the magnetization vector. Without taking into account the spin-polarized charge carrier current, this equation has the form

$$\frac{\mathrm{d}\vec{M}}{\mathrm{d}t} = -\left[\vec{M}, \overrightarrow{H_{\text{eff}}},\right] + \alpha\left[\vec{M}, \frac{\mathrm{d}\vec{M}}{\mathrm{d}t}\right], \tag{1.14}$$

where $\vec{M}$ is the magnetization vector, α is the constant that characterizes the attenuation, and $\overrightarrow{H_{\text{eff}}}$ is the vector of the effective field acting on the magnetization, which is defined as

$$H_{\text{eff}}^{i} = -\frac{\partial W}{\partial S_i}, \quad i = x, y, z, \tag{1.15}$$

where $\vec{S}$ is the spin vector, W is the total magnetic power consisting of the exchange interaction energies, the power of the single-ion anisotropy, the power of the dipole–dipole interaction, and the power of the magnetic moment in an external magnetic field (Zeeman power).

In the absence of attenuation, the magnetic moment, being in the initial state non-collinear with the practical field, makes a precession around the direction of the field (Fig. 1.11a).

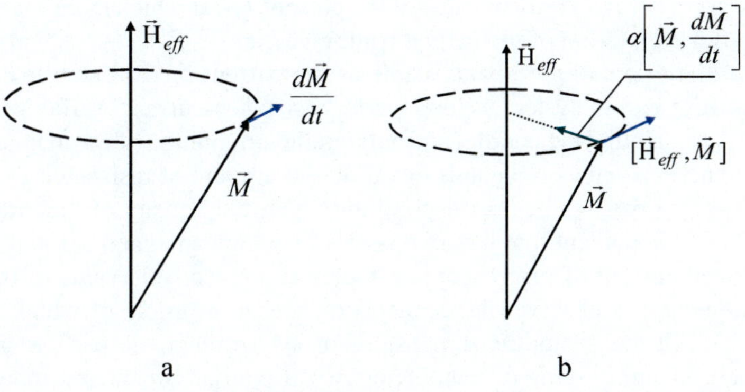

Fig. 1.11 Precession in the absence of (**a**) and in the presence of (**b**) attenuation

The presence of attenuation ($\alpha > 0$) leads to the fact that, over time, the direction of the vector $\vec{M}$ tends to the direction of the effective field (Fig. 1.11b), that is, to the equilibrium state. In this case, the end of the vector $\vec{M}$ describes a spiral.

To describe the action of spin-polarized charge carriers falling into a ferromagnet, the American scientists J. Slonczewski and L. Berger introduced an additional term in the right part of Eq. (1.14) $-\beta\left[\vec{M},\left[\vec{M},\overrightarrow{M_0}\right]\right]$, where $\overrightarrow{M_0}$ is the magnetization vector of the fixed (thick) ferromagnetic layer, and the coefficient β is directly proportional to the strength of the flowing current ($\beta \sim I$) and changes its sign when the direction of the electric current changes.

The solution of the resulting equation is usually found by numerical methods (micro modeling). Depending on the current sign, this term is either codirected with the term $\alpha\left[\vec{M}, \frac{d\vec{M}}{dt}\right]$ or opposite to it in the direction.

In $I < 0$ this case, spin-polarized electrons from a thick ferromagnetic layer fall into a thin ferromagnetic layer. If the current strength exceeds the critical value in modulus, then the magnetization vector of the thin layer, which in the absence of current did not coincide in the direction with the magnetization vector of the thick ferromagnetic layer, will turn so that their principles coincide.

What happens if we change the direction of the current (I > 0)? The spin-polarized electrons, leaving the thin layer, move to the thick layer with a fixed magnetization. The "majority" electrons (with parallel magnetizations of the layers, this term is simultaneously applicable to both layers) are easier to penetrate the thick layer. In contrast, the "minority" electrons are more likely to be reflected from the boundary with the thick layer and return. There is an excess of them compared to the equilibrium state, which increases with increasing current strength. There is a critical value of the recent power (I_{c1}), at which the magnetization vector of a thin ferromagnetic layer will "tip over" and become antiparallel to the magnetization vector of a thick layer [1].

If we begin to reduce the current strength, the antiparallel state will remain stable until the second negative critical value of the present I_{c2}, at which there is a transition to the parallel orientation of the magnetization vectors.

The parallel orientation remains stable as the current increases up to a value of I_{c1}. The corresponding hysteresis loop in the "resistance-current" variables is shown in Fig. 1.12. Recall that the parallel and antiparallel orientation of the magnetizations of ferromagnetic layers corresponds to a different amount of resistance.

The critical current values correspond to the current density of the order of 10^6 A/cm^2. Thus, it is evident that it is impossible to achieve magnetization reversal by spin-polarized current of macroscopic samples at a reasonable value of the winds. This phenomenon is observed in the nanocontact, an example of which is shown in Fig. 1.13. All the geometric dimensions of a ferromagnetic particle in contact have a scale of units—tens of nanometers, so it can be considered monodomain and describe the behavior of its magnetization vector with just one Landau-Lifshitz-Hilbert equation.

It turned out that the behavior of such a nanocontact in an external magnetic field is not limited only to the stationary states considered above. Among the many

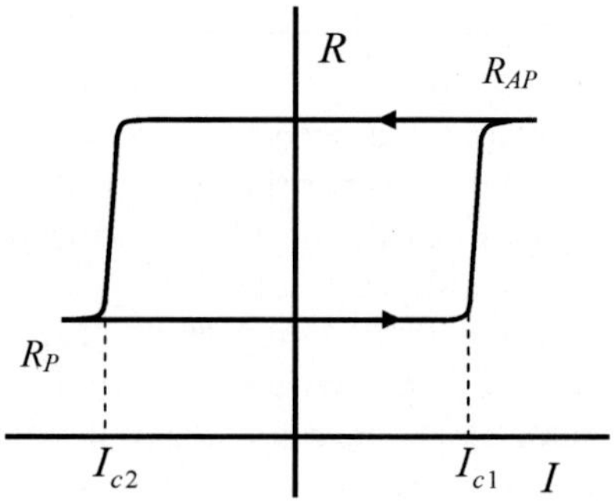

Fig. 1.12 Current hysteresis loop

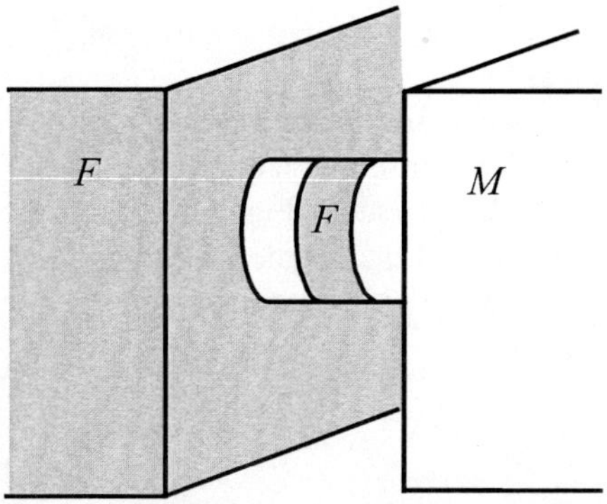

Fig. 1.13 Nanocontact

solutions to the Landau-Lifshitz-Hilbert equation, some solutions correspond to the undamped precession of the magnetization vector of a thin ferromagnetic layer. This precession, caused by the flow of spin-polarized current in an external magnetic field, is accompanied by the generation of electromagnetic waves in the gigahertz frequency range. This generation was first experimentally detected in January 2004. Depending on the current strength, several generation frequencies were observed in the field of 5–40 GHz.

The efficiency of such a microwave radiation source is small, but it creates a unique opportunity to combine a gigahertz radiation source with a microchip. It is necessary to make a series of identical contacts and solve the problem of their auto-phasing to increase the radiation power so that the radiation of the communications occurs in phase, and its intensity increases with the number of contacts N as N^2.

1.5 MRAM (Magnetoresistive Random Access Memory)

Based on GMR and TMR, magnetoresistive memory is created, replacing other types of memory, including fast SRAM, under favorable conditions. It has a quick access time (on the order of units—tens of nanoseconds) and is non-volatile; that is, it is stored in the absence of power supply (unlike existing personal computer EC) implementations. For special applications, it is also essential that it has the property of radiation resistance.

Let us consider the principles of operation of the first MRAM cells based on the effects of GMR and TMR. The design of a memory cell based on a pseudo-spin valve (PSV) is shown in Fig. 1.14.

There are two magnetic layers of materials with different properties, selected so that one layer switches at a lower magnetic field strength, the other—at a higher one. You can use the same material, but the layers are made of different thicknesses. In this case, the thinner film will switch at a lower field strength ("soft" layer), and the thicker film will change at a higher field strength ("hard" layer). The "soft" layer is used to read the information, and the "hard" layer is used to write and store it. The "soft" layer can be repeatedly remagnetized without changing the state of the "hard" layer.

Information is recorded by passing current simultaneously along two lines: the data line (sense line) and the write/read line (word line), at the intersection of which this cell is located (Fig. 1.15). The values "0" and "1" correspond to the opposite directions of the magnetization of the "hard" layer. The write/read line is electrically isolated from the data line.

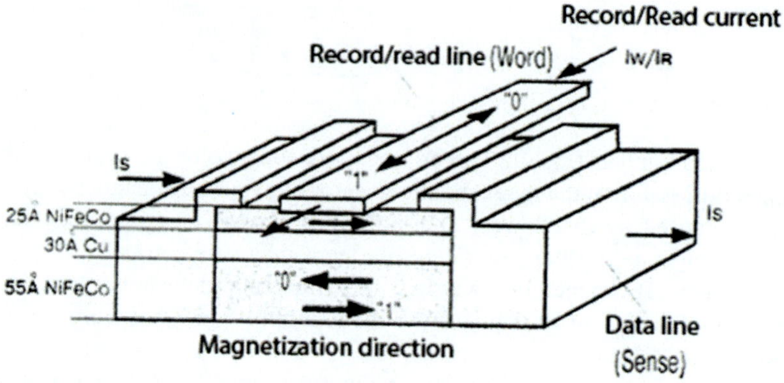

Fig. 1.14 A PSV MRAM memory cell based on the GMR effect

Fig. 1.15 The MRAM geometry

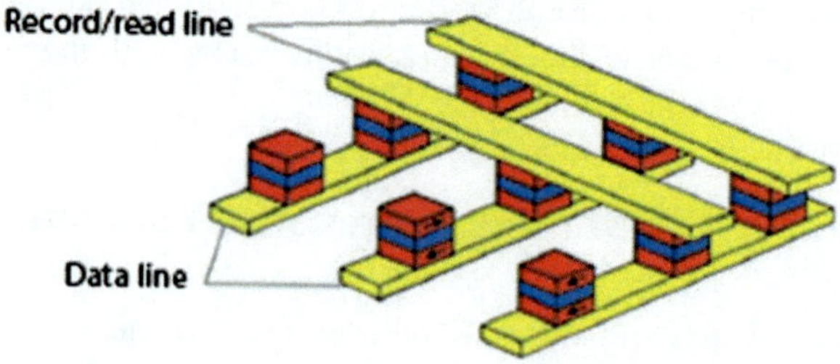

During the reading process, a variable polarity current is passed through the write/read line. The magnetic field created by it is not strong enough to remagnetize the lower layer, but it is sufficient to remagnetize the upper ferromagnetic layer. Therefore, when a current is passed, the parallel orientation of the magnetizations is replaced by an antiparallel direction, etc. If a weak direct current is passed through the data line, then due to the modulation of the resistance of the element located at the intersection of the two lines (Fig. 1.15), the voltage of the data line will also be modulated, but with a lower depth, since the resistance of only one element in the line changes. The state of this particular element is determined during the reading process. Since the voltage on all data lines can be measured simultaneously, all parts located along the write/read line are read simultaneously. Figure 1.16 shows the dependences of the write/read line current, the element resistance, the data line voltage, and its derivative on the same time scale for cases where "0" (a) or "1" (b) is written in the cell.

For reading, it is enough to pass several pulses, and the characteristic reading time is about 50 ns. The recording time is about 100 ns and can be reduced.

The memory cell had a lateral size of 200 nm, the lower magnetic layer had a thickness of 5.5 nm, the upper magnetic layer had a thickness of 2.5 nm, and the middle (copper) layer had a thickness of 3 nm.

In addition, the tests have shown that MRAM can withstand an almost unlimited number of write and read cycles. That is, there are practically no aging and degradation processes.

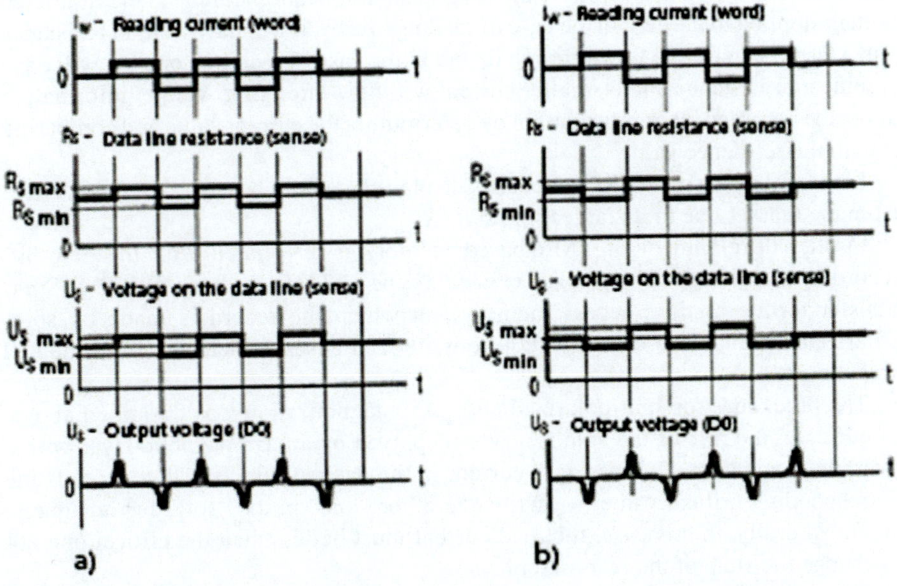

Fig. 1.16 Diagram of the operation of the PSV MRAM when reading "0" (**a**) and "1" (**b**)

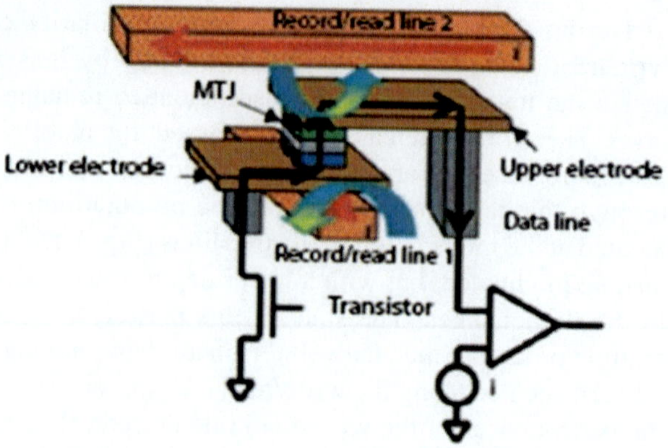

Fig. 1.17 An MRAM cell of type 1 T-1MTJ

The TMR-based memory cell is shown in Fig. 1.17. Each cell contains a transistor (1 T-1MTJ structure). Similar to the case of MRAM based on the GMR effect, a bit of information is written by simultaneously passing current along two write/read lines at the intersection of which this cell is located. In this case, the bit is recorded in the "soft" ferromagnetic layer, and the magnetization of the "hard" ferromagnetic layer remains unchanged.

The reading is performed by applying a voltage to this cell, for which the control transistor is used. Let, for the sake of certainty, in the case of writing "zero", the magnetization of the "soft" layer is parallel to the magnetization of the "hard" layer, and in the case of writing "one", the magnetization is antiparallel. Then, at a fixed voltage applied to the cell, in the case of reading "zero", it will have a slight resistance and a large current will flow through it, and in the case of reading "one", it will have significant resistance and a smaller current will flow through it. We get information about the recorded bit of information by determining the current value and comparing it with the reference value.

Memory based on the TMR effect has displaced memory based on the GMR effect from the market due to its larger size $\Delta R/R$.

Further development of TMR-based memory is associated with the new bit-recording technology. Toshiba has created a type of TMR-memory SPRAM (Spin-transfer torque random access memory), in which the record is made by spin-polarized current. That does not require write/read buses, which simplifies the cell architecture.

The procedure for spin-polarized current remagnetization was described above. In our case, the role of the spin reservoir is played by a ferromagnetic layer with a fixed magnetization. By passing a current in the desired direction that exceeds the corresponding critical value, we write "zero" or "one" in the "soft" ferromagnetic layer. Naturally, in this case, the read current must be less than the critical one not to change the state of the recorded bit.

The transition to SPRAM allowed the write current to be reduced and opened the way for further cell size reduction. The fact is that in traditional MRAM based on the TMR effect, the value of the write current increased with a decrease in the size of the cell inversely proportional to its width. In SPRAM, the write current drops with the cell size.

1.6 Superparamagnetic Limit

Let us consider the fundamental lower bound on the size of monodomain nanoparticles and domains on a hard disk, which is also called the paramagnetic limit.

Let, as a result of the addition of the crystallographic anisotropy and the anisotropy of the shape, the particle (domain) is characterized by uniaxial anisotropy, the volume power density of which has the form—$\tilde{K}M^2\cos^2\theta$, where $\tilde{K} > 0$ is the effective anisotropy constant, and θ is the angle that the magnetization vector $\vec{M}$ is with the easy magnetization axis.

Since the anisotropy power is even in vector $\vec{M}$, the states that differ in the sign $\vec{M}$ have the same power; that is, there are two equivalent minima of the anisotropy power (Fig. 1.18). Therefore, in a state of thermodynamic equilibrium, the system will move from one minimum of power to another. Since the particle is monodomain, this transition occurs by turning the magnetization vector by 180° through the difficult direction ($\theta = 90°$). In this case, the system must overcome a potential barrier U, the value of which is equal to [2]

$$U = W_{\text{an}}\left(\theta = 90^{\circ}\right) - W_{\text{an}}(\theta = 0), \tag{1.16}$$

where W_{an} is the anisotropy power

$$W_{\text{an}} = -\tilde{K}M^2V\cos^2\theta, \tag{1.17}$$

V is the volume of the nanoparticle. From here, we get

$$U = \tilde{K}M^2V. \tag{1.18}$$

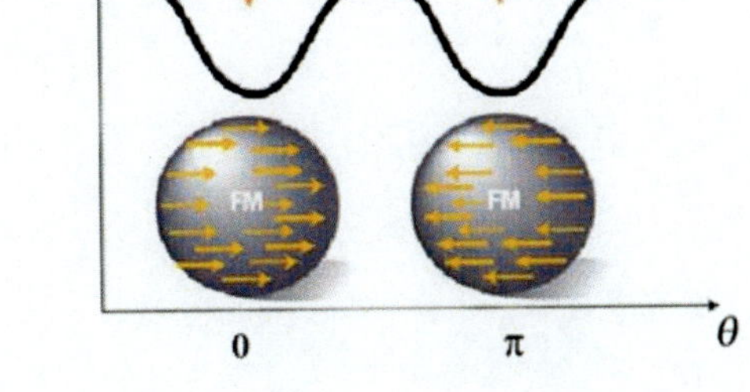

Fig. 1.18 A power barrier separating two equilibrium states of a particle with anisotropy of the "light axis" type

It is assumed that at $U \geq 70T$ (T—temperature, measured in power units) during the memory operation, the probability of such reorientation processes is negligible, and the direction $\vec{M}$ in the absence of external influence (rewriting) remains unchanged. Thus, we get a limit on the volume of the nanoparticle (domain):

$$V \geq \frac{70T}{\tilde{K} M^2}. \tag{1.19}$$

To reduce the limit size of the nanoparticle, it is necessary to find substances with a significant value of the anisotropy constant and the magnetization value. It should be noted that such materials will have a higher value of the coercive field, since the same power barrier must be overcome during the magnetization reversal U. Large currents will be required to write information to such media by a magnetic field. Therefore, the search for such materials should be accompanied by searching for alternative ways of recording information (thermally activated recording, electric field recording).

References

1. Morosov AI, Sigov AS (2004) A new type of domain walls–domain walls generated by frustrations in multilayer magnetic nanostructures (Review). FTT 46(3):385–400
2. Morosov AI, Sigov AS (2012) Frustrated multilayer ferromagnet-antiferromagnet structures: going beyond the exchange approximation (Review). FTT 54(2):209–229

Chapter 2
Frustrations of Exchange Interaction

2.1 Why Does Spin Feel Hopeless?

The term "frustration" came to physics from physiology. The latter defines frustration [lat. frūstrātio-deception; failure, failure; breakdown] as a psychological state that occurs in a situation of frustration, failure to fulfill some significant goal or need for a person and manifests itself in oppressive tension, anxiety, and a sense of hopelessness. What does this metaphor mean concerning the spin?

Let us consider a system of three spins localized at the vertices of an equilateral triangle (Fig. 2.1), between which the Heisenberg exchange interaction takes place [1]

$$\hat{H}_{\text{ex}} = -J \sum_{\substack{i=1 \\ j>1}}^{3} \left(\widehat{S_i}, \widehat{S_j}\right), \tag{2.1}$$

where J is the exchange integral, $\widehat{S_i}$ is the spin operator, and the indices i, j number the specified spins.

In the case $J < 0$ of (antiferromagnetic exchange interaction), the interaction power of each pair of spins is minimal when they are oriented antiparallel to each other. However, it is impossible to find an orientation for a triple of spins in which the spins of any pair are antiparallel to each other. Indeed, if spins 2 and 3 are antiparallel to spin 1, they are oriented parallel to each other (Fig. 2.1). The impossibility of the existence of a spin orientation for which the power of the exchange interaction of any pair of spins reaches a minimum, we will call the frustration of the exchange interaction of spins. Similarly, we can introduce the concept of frustration for another type or sum of interactions.

A. S. Sigov, *Multilayer Magnetic Nanostructures*, Springer Aerospace Technology,
https://doi.org/10.1007/978-981-19-6246-2_2

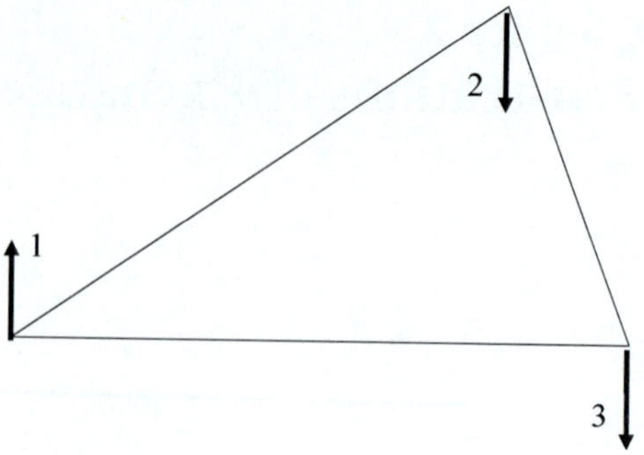

Fig. 2.1 The frustrated three-spin system with antiferromagnetic exchange interaction

2.2 Frustrations in a System with a Non-magnetic Layer

The simple picture of the magnetic ordering in a spin-valve system with a non-magnetic layer, given in the previous chapter, is valid within the framework of the model assumption of the ideal smoothness of the interfaces between the layers. However, in reality, the boundaries of the layers are rough; that is, the thickness of the coating is different in different areas. In crystalline layers, which we will limit ourselves to, the roughness is due to the presence of atomic steps at the interface (Fig. 2.2), which change the thickness of the layer by one monoatomic layer.

To understand what happens with the interaction between the ferromagnetic layers which is the sum of the paired RKKY interactions of the spins belonging to different layers, it is necessary to solve the question of which area contributes to the molecular field acting from the second layer on the spin of the atom belonging to the first layer.

A simple analysis shows that this area is located directly opposite the atom and has a characteristic size of the order of the thickness of the non-magnetic layer d (Fig. 2.2). In other words, the parameter d characterizes the non-locality of the interaction of layers: the exchange interaction between layers at a given point in the layer plane is determined not by the local value of the layer thickness, but by its weight in some area having a longitudinal (parallel to the layer plane) size of order d.

We will assume that d is smaller than the other scales that arise in the problem of magnetic ordering in the structure under consideration; that is, we will neglect

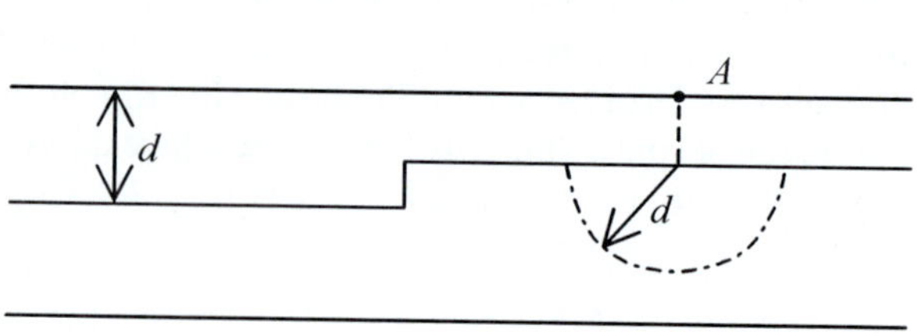

Fig. 2.2 The atomic step at the interface of the layers. The dotted line shows the area that makes the main contribution to the molecular field at point A

non-locality, assuming that $J_\perp(x, y) = J_\perp(d(x, y))$ and is described by the formula (1.1), where the x- and y-axes of the Cartesian rectangular coordinate system lie in the plane of the layers, the z-axis is perpendicular to it, and the boundary of the stage is parallel to the y-axis.

If on one side of the step $J_\perp = J_\perp(n)$, then on the other $J_\perp = J_\perp(n-1)$. In the case when $J_\perp(n)J_\perp(n-1) < 0$, we will say that there is frustration in the system. Indeed, the homogeneous distribution of the order parameters (in this case, the magnetization) in the layers, which corresponds to the minimum of the exchange power in each layer, does not reach the minimum of the interaction power between the layers. If this power is minimal on one side of the atomic step, then it is maximal on the other side.

The question of what state is realized in this frustrating system will be discussed in the next chapter.

2.3 Frustrations in the Ferromagnet–Antiferromagnet System

Let us now consider the concept of the frustration of the exchange interaction at the interface of the ferromagnet–antiferromagnet layers. Let us consider two cases [2]:

1. The atomic planes of a two-lattice antiferromagnet parallel to the interface contain the spins of only one sublattice, and the spins of neighboring atomic planes belong to different sublattices. We will call such a cross-section of the antiferromagnet surface uncompensated.
2. The atomic planes of the antiferromagnet parallel to the interface contain an equal number of spins belonging to different sublattices, and in the absence of external influence and the Dzyaloshinskii-Moriya interaction, they have zero magnetic moments. We will call such a cross-section of the antiferromagnet surface compensated.

2.3.1 Uncompensated Cross-Section of the Antiferromagnet Surface

In the future, we will need to solve a rather complex two-dimensional problem about the inhomogeneous non-collinear distribution of spins so that we will use a simple model of the Heisenberg exchange interaction of quasi-classical localized spins in the approximation of the interaction between the nearest neighbors. We will assume that the crystal lattice is a continuation of the volume-centered tetragonal or cubic lattice of a two-lattice collinear antiferromagnet. The difference in the lattice constants is negligible.

In addition, we restrict ourselves to the temperature range $T \ll T_C, T_N$, where T_C is the Curie temperature of the ferromagnet, and T_N is the Neel temperature of the antiferromagnet. In this domain, we can assume that the values of the localized spins remain unchanged, but only their orientation changes, and characterize i the localized spin by the unit vector $\mathbf{s}_i$, assuming that the spin modulus is included in the corresponding interaction constants: the exchange integral and the anisotropy constant.

Taking into account these remarks, the power of the exchange interaction of spins is written as

$$W_{ij}^{\mathrm{ex}} = -J_{ij}\left(\widehat{S}_i, \widehat{S}_j\right) \tag{2.2}$$

where $J_{ij} = J_{\mathrm{f}} > 0$, if the neighboring spins belong to a ferromagnet, $J_{ij} = J_{\mathrm{af}} < 0$, if they belong to an antiferromagnet, $J_{ij} = J_{\mathrm{f,\,af}}$, if the neighboring spins belong to different layers. For the sake of certainty, we will assume that $J_{\mathrm{f,\,af}} > 0$.

Consider a cross-section plane (100) of a tetragonal or cubic volume-centered crystal lattice of an antiferromagnet.

There is no frustration of the exchange interaction in the case of a smooth surface of the "ferromagnetic–antiferromagnetic" interface. If $J_{\mathrm{f,\,af}} > 0$, the spins of the ferromagnet in the state with minimal power are directed parallel to the spins of the upper uncompensated atomic plane of the antiferromagnet (Fig. 2.3a). The $J_{\mathrm{f,\,af}} < 0$ orientation would be antiparallel.

However, the actual interface contains steps that change the layer thickness by one atomic plane. In principle, actions with a height of several atomic planes are also possible. If the number of these planes is even, then such steps do not lead to frustration. If the number of these atomic planes is odd, then the frustration caused by such a step is equivalent to the frustration created by an atomic step with a height of one atomic plane, which we will limit ourselves to considering.

On different sides of the atomic step, the spins of the ferromagnet are adjacent to the spins of different sublattices of the antiferromagnet. If we assume that the ferromagnetic order parameter—the magnetization of the ferromagnet—and the antiferromagnetic oder parameter—the difference in the magnetizations of the sublattices of the antiferromagnet—are distributed uniformly in their layers, then on one side of the step, the mutual orientation of the ferro- and antiferromagnetic order parameters will correspond to the minimum power of the exchange interaction between the layers, and on the other side of the step—the maximum of this power (Fig. 2.3b). Thus, by choosing a homogeneous distribution of the order parameters corresponding to the minimum of the exchange interaction in the layers, we cannot minimize the power of the interlayer exchange interaction. Consequently, the presence of an atomic step at the interface leads to frustration.

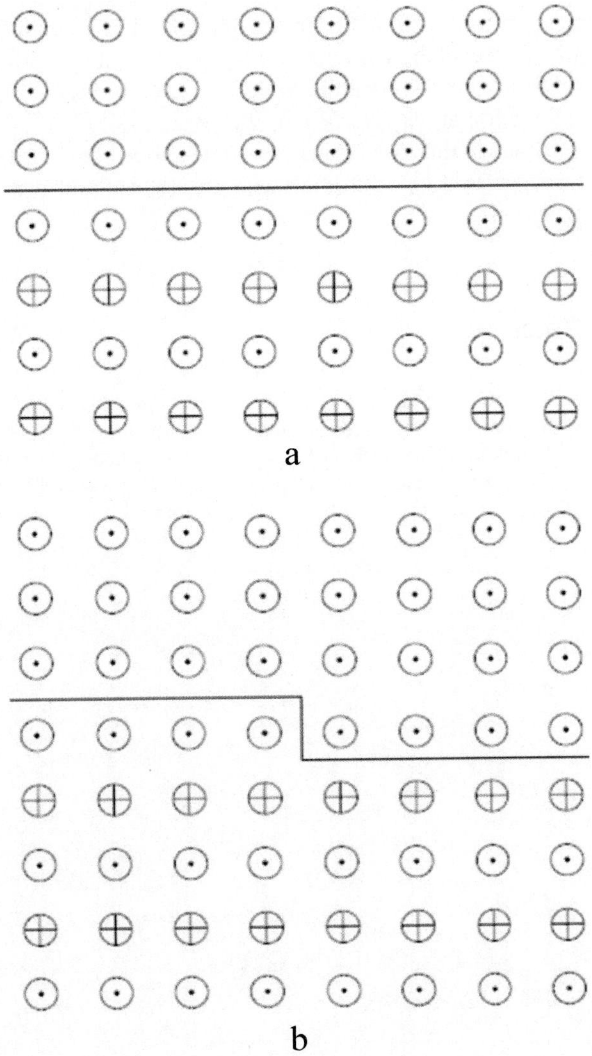

Fig. 2.3 Ideally flat (**a**) and containing an atomic step (**b**) of the interface between a ferromagnet and an uncompensated antiferromagnet

2.3.2 *Compensated Cross-Section of the Antiferromagnet Surface*

Let us now consider the compensated slice (110) of an antiferromagnet's tetragonal or cubic volume-centered lattice. Frustration occurs on a compensated slice, even in the case of an atomically smooth interface. Indeed, the exchange interaction between the spins of a ferromagnet tends to orient them parallel to each other. The exchange interaction between the spins of the lower plane of the ferromagnet and the spins of the upper plane of the antiferromagnet tends to orient the spins of the antiferromagnet parallel to the spins of the ferromagnet, that is, parallel to each other. At the same time, the exchange interaction between the neighboring spins of the antiferromagnet tends to orient them antiparallel to each other. There is a frustration of the exchange interaction at the interface of the layers.

Since in the case of a compensated slice of an antiferromagnet, frustration occurs already for a smooth interface, atomic steps on it do not significantly change the situation.

References

1. Morosov AI, Sigov AS (2004) A new type of domain walls—domain walls generated by frustrations in multilayer magnetic nanostructures (Review). FTT 46(3):385–400
2. Morosov AI, Sigov AS (2012) Frustrated multilayer ferromagnet-antiferromagnet structures: going beyond the exchange approximation (Review). FTT 54(2):209–229

Chapter 3
Domain Walls and the Phase Diagram of the Spin-Valve Systems with a Non-magnetic Layer

3.1 Domain Wall Generated by Frustration

Let us consider a rectilinear solitary stage on one of the interfaces of a three-layer system (Fig. 2.2). Let for certainty

$$J_{\perp}(x) = \begin{cases} J_1 > 0, & \text{when } x < 0 \\ J_1 < 0, & \text{when } x > 0 \end{cases} \tag{3.1}$$

It is pretty evident that, far from the stage, the relative orientation of the magnetizations of the layers must satisfy the condition of the minimum interaction power; that is, the magnetizations must be parallel at $x \to -\infty$ and antiparallel at $x \to +\infty$.

A domain boundary must arise near the edge of the stage, which permeates all the layers and separates the half-space with a parallel orientation of the magnetizations of the layers ($x < 0$) from the half-space with an antiparallel orientation of the magnetizations ($x > 0$) to do this.

Here and in the future, we will not touch on mixing materials of neighboring layers. If this effect is weak and does not mask the presence of steps at the boundaries of the layers, then it can be taken into account by renormalizing the constant of the interlayer exchange interaction. In the case of solid mixing of the materials of the layers, the very concept of the interface becomes indeterminate. It should be noted, however, that such imperfect structures are not widely used in practice.

We will assume that the spins of the atoms lie in the plane of the layers, and therefore, in the case of perfectly smooth interfaces, there are no scattering fields. The angle gives the position of the I th spin θ_i that the spin forms with the *x-axis*. The modulus of the order parameter will be considered unchanged.

To begin with, we consider the resulting domain wall in the exchange approximation; we neglect the anisotropy power in the plane of the layers. This is true if the power of the frustration that causes the appearance of the domain wall is much higher than the power of the anisotropy, which is true if the width of the resulting

A. S. Sigov, *Multilayer Magnetic Nanostructures*, Springer Aerospace Technology,
https://doi.org/10.1007/978-981-19-6246-2_3

domain wall is much smaller than the width of the traditional domain wall, which is determined by the competition of the exchange interaction and anisotropy energies.

The calculation will be carried out within the framework of the continuum approximation. As will be shown below, the characteristic width of the domain wall far exceeds the thickness of the layers in the nanostructure, so it can be assumed that it does not depend on the coordinate perpendicular to the layers. Then, in the case of a rectilinear step edge, the problem becomes one-dimensional.

The addition to the power of the exchange interaction of spins in the layers due to the inhomogeneity of the order parameter (magnetization) per unit length of the step can be represented as [1]

$$\tilde{w}_1 = \int\left[\frac{\alpha_1}{2}(\theta_1')^2 + \frac{\alpha_2}{2}(\theta_2')^2\right]dx, \tag{3.2}$$

where θ_i is the angle of rotation of the order parameter in the i th ferromagnetic layer ($i = 1, 2$), the stroke means a differentiation in x. In order of magnitude, the exchange stiffness of layers α_i equal to

$$\alpha_i \sim J_i S_i^2 l_i / b, \tag{3.3}$$

where J_i is the power of the exchange interaction between neighboring spins in *the* i th layer, S_i is the average value of the spin of the atom, l_i is the thickness of the i th layer, and b is the interatomic distance, which for simplicity, we will assume the same for all layers.

The interaction power of the layers in the mean-field approximation has the form

$$\tilde{w}_2 = \int \beta(x)\cos(\theta_1 - \theta_2)dx, \tag{3.4}$$

where

$$\beta(x) = \begin{cases} \beta_1 > 0, & \text{when } x < 0, \\ -\beta_2 < 0, & \text{when } x > 0, \end{cases} \sim J_\perp(x) S_1 S_2 b^{-2}. \tag{3.5}$$

Varying the sum $\tilde{w}_1$ of $+\ \tilde{w}_2$ by θ_1 and θ_2, we obtain the following system of equations:

$$\alpha_1\theta_1'' - \beta\sin(\theta_1 - \theta_2) = 0,\ \alpha_2\theta_2'' + \beta\sin(\theta_1 - \theta_2) = 0, \tag{3.6}$$

with boundary conditions $\theta_i' \to 0$ at $x \to \pm\infty$; $\theta_i \to 0$ at $x \to -\infty$, $|\theta_1 - \theta_2| \to \pi$ at $x \to +\infty$. The solution of this equation has the form $\theta_2 = -\alpha_1\theta/(\alpha_1 + \alpha_2)$; $\theta_1 = \alpha_2\theta/(\alpha_1 + \alpha_2)$, and the equations give the dependence

$$\cos\frac{\theta}{2} = -th\left[\left(\frac{\beta_1}{\alpha^*}\right)^{1/2}(x + x_1)\right],\ x < 0;$$

$$\sin\frac{\theta}{2} = th\left[\left(\frac{\beta_2}{\alpha^*}\right)^{1/2}(x + x_2)\right], x > 0, \tag{3.7}$$

where $\alpha^* = \alpha_1\alpha_2/(\alpha_1 + \alpha_2)$, and the constants x_1 and x_2 are found from the condition of continuity $\theta(x)$ and its derivative $\theta'(x)$ at $x = 0$, which leads to the equation

$$tg\left.\frac{\theta}{2}\right|_{x=0} = \left(\frac{\beta_2}{\beta_1}\right)^{1/2}. \tag{3.8}$$

It is easy to see that at $\beta_1 >> \beta_2$ almost the entire wall is located in the area $x > 0$, and at $\beta_1 << \beta_2$-in the area $x < 0$. In the case $\alpha_1 = \alpha_2$ of the back of the layers, they turn in different directions at an angle of 90°. If one of the values α is much higher than the other, which corresponds to a much larger thickness of one of the layers, then almost the entire spin rotation occurs in the thinner layer, and in the thick layer, the spins deviate by a slight angle.

The characteristic width of the domain wall δ is [2]

$$\delta = \pi\left(\frac{\alpha^*}{\min(\beta_1, \beta_2)}\right)^{1/2} \sim \pi b\left(\frac{J_i l_{\min}}{J_\perp b}\right)^{1/2} \sim \pi d\left(\frac{J_i l_{\min}}{J_0 b}\right)^{1/2}, \tag{3.9}$$

where $l_{\min}$ is the smaller of the thicknesses of the two magnetic layers, and from $J_\perp$ we have isolated the dependence on the thickness of the interlayer: $J_\perp \equiv J_0(b/d)^2\sin(2k_F d)$. For $l/b \sim 3$–5, $J_i/J_0 \sim 1$–10, and $d \sim 1$ nm, we get $\delta \sim 10$ nm, much less than the width of the domain wall in iron (80 nm).

The value δ increases with the thickness of the nanostructure layers. If it becomes comparable to the width of a traditional domain wall, then the anisotropy power must be taken into account. With a further increase in the thickness of the layers, the considered domain wall is transformed into a traditional Neel or combined domain wall.

Substituting the solution (3.7) in the functional $\tilde{w}_1 + \tilde{w}_2$, we obtain the power of the domain wall integrated over the thickness of the layers, that is, the power per unit length of the domain line on the surface of the layers as the difference between this power and the sum of the energies of homogeneous states with $\theta = 0$ at $x < 0$ and $\theta = \pi$ at $x > 0$,

$$\begin{aligned}\sigma &= 4(\alpha^*)^{1/2}[\beta_1^{1/2} + \beta_2^{1/2} - (\beta_1 + \beta_2)^{1/2}] \\ &\sim d^{-1}S^2[J_i J_0 l_{\min}/b]^{1/2} \sim b^{-1}S^2[J_i J_\perp l_{\min}/b]^{1/2}.\end{aligned} \tag{3.10}$$

3.2 Phase Diagram

We now study the phase diagram of the spin-valve structure as a function of the thickness of the layers and the degree of roughness of their interfaces. We restrict ourselves to the case *of* $R \gg d$ to still use the local approximation for $J_{\perp}(x, y)$ in the case *of* $R \ll d$, the value is effectively averaged $J_{\perp}(x, y)$ over the area of nonlocality, and the system becomes equivalent to the issue of smooth interfaces, but with an averaged value $J_{\perp}$.

If the characteristic distance R between the atomic steps leading to frustration at the interfaces of the layers is much greater than the width of the domain wall $\delta(R \gg \delta)$, then it is energetically advantageous for the magnetic layers to split into domains. The domain walls separate the areas with parallel and antiparallel orientation of the magnetizations of the ferromagnetic layers. The picture of the partition into domains is shown in Fig. 3.1 for the case of identical thicknesses of ferromagnetic layers. The typical size of such domains can be tens to hundreds of nanometers.

It is valid if the characteristic scale R is much smaller than the width of the traditional domain wall in the ferromagnetic layer Δ_{f} ($\Delta_{\mathrm{f}} \gg R \gg \delta$), and also if $\Delta_{\mathrm{f}} \ll R$ and in the plane of the layer there are two mutually perpendicular easy axes. Let the left-hand magnetization steps of the ferromagnetic layers be parallel to the easy axis. Then, after turning in the domain wall of each magnetization at an angle of 90°, they again turn out to be collinear to the easy axis (already different).

If there is only one easy axis in the plane of the layer, then after such a rotation, the magnetizations are collinear to the "difficult" direction. Therefore, in this case, it is necessary to consider the anisotropy power and conduct a more thorough analysis

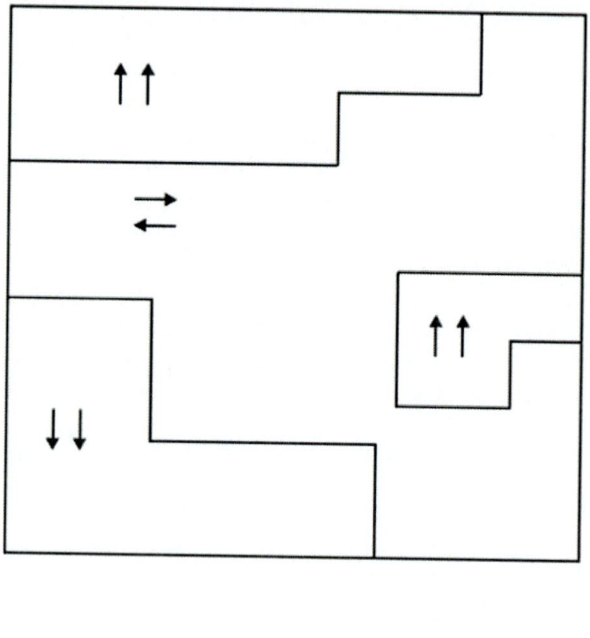

Fig. 3.1 Picture of the partition of the three-layer structure into domains with parallel and antiparallel orientation of the magnetizations of the ferromagnetic layers

of the situation. It shows that the domain wall structure changes dramatically: it now consists of three areas if you move along the x coordinate.

In the first area, both magnetization vectors are rotated in the plane of the layer at an angle of 45° from the easy axis. The thickness of this area is about the width of the traditional domain wall Δ_f, and with this reversal, the magnetizations of the ferromagnetic layers remain parallel to each other.

The second area is the domain wall described above, generated by frustration. Its width is given by the formula (3.9), in which the magnetization vector of one ferromagnetic layer is rotated 90°, preserving the direction of rotation from the first area, and the magnetization vector of the second layer is rotated 90° in the opposite direction. As a result, the magnetizations will turn from the original order at angles of 135° and −45° respectively (Fig. 3.2), and their orientation will become antiparallel.

In the last, third area, the magnetization vectors, while remaining antiparallel, turn at an angle of 45° in the same direction as in the first area. At the exit from the wall, they form 180° and 0° tips with an "easy axis", respectively.

If $\Delta_f \gg R \gg \delta$, then the first and third areas cannot be formed on the domain size, and the exchange approximation is valid.

As the layer thicknesses increase, the "three-layer" domain walls transform. The width of the central area of the wall is compared with the widths of the other areas, and the domain wall in the second ferromagnetic layer disappears. The domain wall in the first ferromagnetic layer is transformed into a traditional 180° domain wall [3].

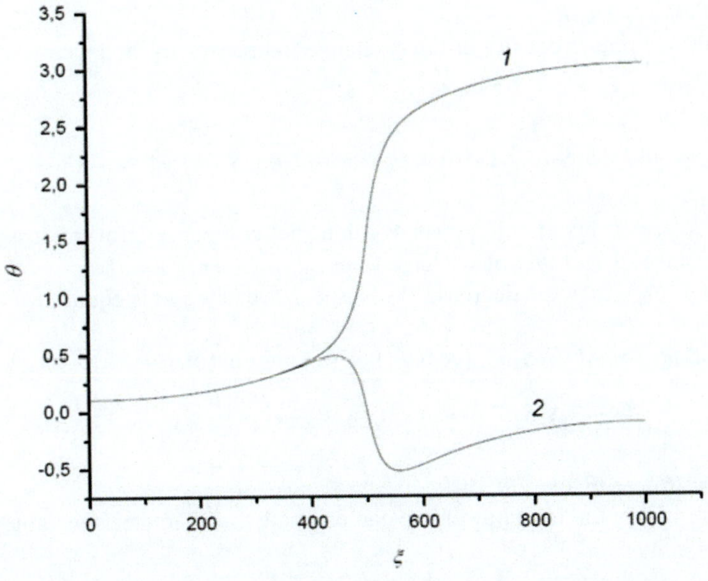

Fig. 3.2 Turn angles of the magnetizations of ferromagnetic layers 1 and 2 in the domain wall consisting of three areas. The edge of the atomic step is located at the point with the coordinate $\zeta = 500$

Suppose the characteristic roughness scale is $R \ll \delta$, then there is not enough space even to form the central area of the domain wall. That is the area of the weak distortion of the magnetic order parameter. The transition from the state $R \ll \delta$ to $R \gg \delta$ can occur as the thickness of the interlayer increases, since $\delta \propto d$. In the area $R \ll \delta$, in principle, only slight deviations $\psi_i(x, y) = \theta_i(x, y) - \langle\theta_i\rangle$ of the angles θ_i from their average values are possible $\langle\theta_i\rangle(|\psi_i| \ll 1, i = 1, 2)$.

We demonstrate the power disadvantage of such deviations $\langle\theta_1\rangle = \langle\theta_2\rangle$.

By varying the total power of the layers $W_1 + W_2$ by ψ_i, it is easy to show that ψ_1 and ψ_2 are expressed in terms of a variable $\psi = \psi_1 - \psi_2$ as follows

$$\psi_1 = \frac{\alpha^*}{\alpha_1}\psi, \psi_2 = -\frac{\alpha^*}{\alpha_2}\psi. \tag{3.11}$$

$$W_1 = \frac{\alpha^*}{2} \int (\nabla\psi)^2 \mathrm{d}S, \tag{3.12}$$

integration takes place over the layer plane.

Due to the appearance of inhomogeneity with a characteristic maximum deviation of the magnetization vector by an angle ψ_0, the surface power density of the layer $w_1 = W_1/\sigma$ (σ—the area of the layers) increases by [2]

$$\Delta w_1 \approx \alpha^* \left(\frac{\psi_0}{R}\right)^2, \tag{3.13}$$

because $|\nabla\psi| \approx \psi_0/R$.

The surface power density of the exchange interaction of the layers $w_2 = W_2/\sigma$ changes by

$$\Delta w_2 \approx -|J_\perp| S_1 S_2 b^{-2}(1 - \cos\psi_0) \approx -|J_\perp| S_1 S_2 b^{-2}\psi_0^2 \approx -\alpha^* \frac{\psi_0^2}{\delta^2}. \tag{3.14}$$

Since $R << \delta$ this gain in power W_2 does not compensate for the loss in power W_1, in the case $\langle\theta_1\rangle = \langle\theta_2\rangle$ of we have $\psi_1(x, y) = \psi_2(x, y) = 0$.

If $\langle\theta_1\rangle \neq \langle\theta_2\rangle$, then the decrease W_2 is linear over ψ_0 and is equal to

$$\begin{aligned}\Delta w_2 &\approx -J_\perp S_1 S_2 b^{-2}[\cos(\langle\theta_1\rangle - \langle\theta_2\rangle) - \cos(\langle\theta_1\rangle - \langle\theta_2\rangle + \psi_0)] \\ &\approx -J_\perp S_1 S_2 b^{-2} \sin\theta \sin\psi_0 \approx -\alpha^* \frac{\sin\theta}{\delta^2}\psi_0\end{aligned} \tag{3.15}$$

where $\theta = \langle\theta_1\rangle - \langle\theta_2\rangle$.

By minimizing the total power, we can estimate the characteristic value of ψ_0

$$\psi_0 \approx \frac{R^2}{\delta^2} \sin\theta, \tag{3.16}$$

as well as reducing the total power

$$\Delta w_1 + \Delta w_2 \approx \frac{\langle J_\perp^2 \rangle S_1^2 S_2^2 \sin^2\theta R^2}{\alpha^* b^4} \approx -\frac{\langle J_\perp^2 \rangle \sin^2\theta R^2}{J_0 l_{\min} b^3} \tag{3.17}$$

where $l_{\min}$ is the minimum of the thicknesses l_1 and l_2 magnetic layers.

Thus, to obtain a gain in power due to the occurrence of weak inhomogeneities of magnetization within the magnetic layers, it is necessary to have a non-collinear state with $\theta \neq 0$.

But the occurrence of such a state is accompanied by a loss in the power of a homogeneous state due to the summand

$$w_2^{(0)} = -\langle J_\perp \rangle S_1 S_2 b^{-2} \cos\theta. \tag{3.18}$$

In the phenomenological description, it is interpreted as the surface power density of bilinear exchange. Replacing $\sin^2\theta$ in the expression (3.17) by $1 - \cos^2\theta$, the term, proportional $\cos^2\theta$, is interpreted as the surface power density of the biquadratic exchange

$$w_{\mathrm{BQ}} = -J_{\mathrm{BQ}} S_1^2 S_2^2 b^{-2} \cos^2\theta. \tag{3.19}$$

The form of the exchange integral J_{BQ} for the case of periodically arranged steps was first defined in [J. C. Slonczewski. Phys. Phys. Lett. **67**, 3172 (1991).]. Note that the value of the exchange integral J_{BQ} is always negative; that is, the biquadratic exchange contributes to the appearance of a non-collinear state. In order of magnitude

$$J_{\mathrm{BQ}} \approx -\frac{\langle J_\perp^2 \rangle R^2}{J_i S^2 l_{\min} b} \tag{3.20}$$

For a non-collinear ordering to occur $\theta \neq 0$, the condition must be met

$$|\langle J_\perp \rangle| \langle 2 | J_{\mathrm{BQ}} | S_1 S_2. \tag{3.21}$$

Assuming that $|J_1| \approx |J_2| \approx \langle J_\perp^2 \rangle^{1/2}$, we obtain, up to numerical multipliers of the order of one, that the inequality (3.21) is equivalent to the condition

$$\frac{\langle J_\perp \rangle}{\langle J_\perp^2 \rangle^{1/2}} < \frac{R^2}{\delta^2} \tag{3.22}$$

It seems unlikely that the values J_1 and J_2 the total areas occupied by the sections with the thickness of the interlayer d_1 and d_2, respectively, are correlated so that

$$\frac{\langle J_\perp \rangle}{\langle J_\perp^2 \rangle^{1/2}} < 10^{-2}.$$

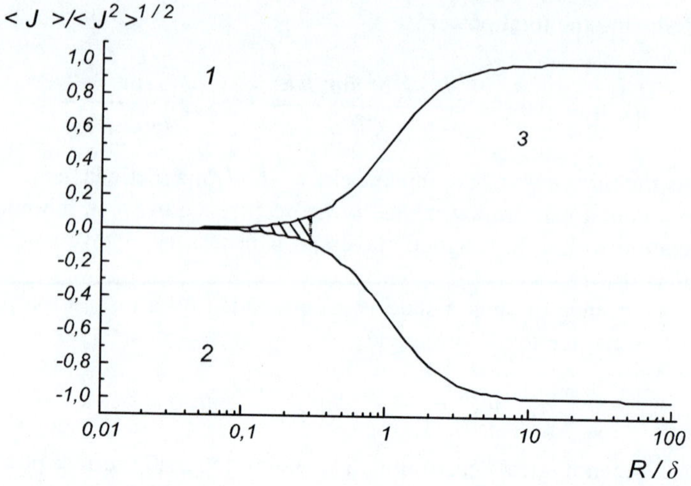

Fig. 3.3 Exchange-roughness phase diagram for a three-layer system with a non-magnetic layer. 1—phase with the parallel orientation of magnetizations of ferromagnetic layers, 2—phase with the antiparallel direction of magnetizations, 3—polydomain phase. The area of existence of a non-collinear state is shaded

Therefore $R << \delta$, only collinear ordering should occur in multilayer structures.

The resulting exchange-roughness phase diagram is shown in Fig. 3.3. The area of the non-collinear phase is shaded, in which the magnetizations of the layers can be considered almost uniform.

It is easy to see that the range of parameter values for which this approximation, which is often found in the literature, is adequate, is unusually narrow. As the size increases R (the roughness decreases), the magnitude of the angle ψ_0 of deviation of the magnetization vector from its average direction increases, and a continuous transition to the nanodomain state occurs [3].

3.3 Behavior in a Magnetic Field

Consider the behavior of different phases in a magnetic field applied parallel to the plane of the layers, considering the anisotropy in the plane of the layers to be negligible.

A narrow rectangular hysteresis loop will be observed in the area with parallel orientation of the magnetizations of the ferromagnetic layers (Fig. 3.4a).

In the area with the antiparallel direction of the magnetizations of two identical ferromagnetic layers, there will be a smooth increase in the total magnetization (Fig. 3.4b), which is entirely analogous to the behavior of a two-sublattice mirror antiferromagnet with an intersublattice exchange power equal to β. The angle between the magnetizations of the layers is found from the minimum power

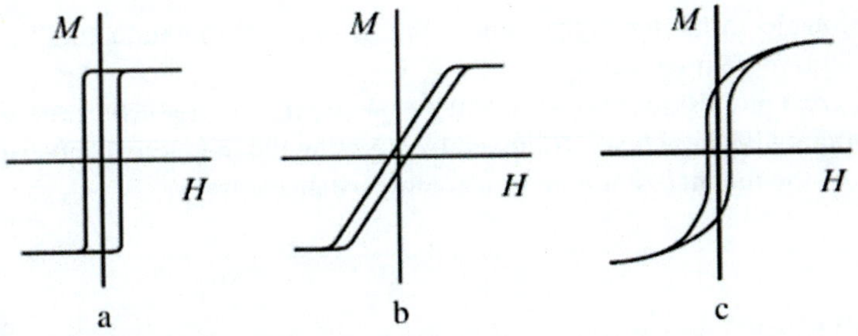

Fig. 3.4 View of the hysteresis loops in different areas of the phase diagram: in the areas with parallel (**a**) and antiparallel (**b**) orientations of the magnetization of ferromagnetic layers and the area of the existence of the nanodomain state (**c**)

$$\tilde{W} = -2M_0 l B \cos\theta - \beta \cos 2\theta, \tag{3.23}$$

where M_0 is the magnetization of the ferromagnetic layers, l is their thickness, $\beta < 0$, and θ is the angle between the magnetic induction and the magnetic moment of the ferromagnetic layer.

It is easy to see that saturation takes place in the field

$$B_{\text{sat}} = \frac{2|\beta|}{M_0 l}. \tag{3.24}$$

In the area of the existence of the nanodomain state, the magnetization curve will be, in the first approximation, a superposition of the two previous curves with weights reflecting the proportion of areas with parallel and antiparallel orientation of the magnetization (Fig. 3.4b).

The observation of such a magnetization curve for the structure under consideration indicates a high degree of probability of a nanodomain state. The only exception is the small shaded area in Fig. 3.3.

3.4 Experimental Observations

Hysteresis loops similar to the one shown in Fig. 3.4b have been observed in a large number of studies; see, for example, [A. Dinia, S. Zoll, M. Gester et al. Eur. Phys. J. B **5**, 203 (1998); C. H. Marrows, J. Hickey, M. Herrman et al. Phys. Phys. B **61**, 4131 (2000)]. However, they were interpreted in terms of biquadratic exchange, and the magnetic structure of ferromagnetic layers on nanoscales, as a rule, was not studied. Nanodomains with parallel and antiparallel orientation of the layer magnetizations were observed by spin-polarization low-power electron microscopy (SPLEEM) [E. Bauer, T. Duden, H. Pinkvos et al. J. Magn. Magn. Mater. **156**, 1 (1996)]. At the same time, it was found that the boundaries of the nanodomains coincide with the boundaries of the atomic terraces at the interface. Nanodomains can also be observed by transmission electron microscopy, spin-polarization scanning tunneling microscopy, and magnetic force microscopy.

It is of great interest to conduct in situ measurements of the surface relief of layers before spraying the next layer, followed by studying the magnetic nanostructure and the study of the magnetization and magnetoresistance curves.

References

1. Morosov AI, Sigov AS (1995) The surface roughness of the layers and the giant magnetoresistance of magnetic multilayer structures. Lett JETF 61(11):893–898
2. Morosov AI, Sigov AS (1996) Giant magnetoresistance of magnetic multilayer structures in the case of weak roughness. FTT 38(9):2823–2830
3. Morosov AI, Sigov AS (1997) The roughness of the layer interface and the phase diagram of magnetic multilayer structures. FTT 39(7):1244–1247

Chapter 4
A Thin Film of a Ferromagnet on an Antiferromagnetic Substrate: Uncompensated Slice

4.1 Model Description

We assume that the antiferromagnetic parameter $\vec{L}$ equal to the difference in the magnetization of the sublattices lies in the plane of the layers and can, as in the previous section, be described by the angle that the vector $\overrightarrow{L}$ forms with the x-axis$\left(\left|\vec{L}\right| = \text{const}\right)$

Here, in contrast to the previous section, we cannot limit ourselves to a one-dimensional problem since the widths of the domain walls in the system under consideration, as will be shown below, are comparable to the atomic scale and vary significantly with the distance from the interface of the layers. Therefore, even in a rectilinear and boundless edge of the atomic step, it will be necessary to solve a two-dimensional problem [1].

Consider a section (100) of a tetragonal volume-centered lattice, with such a ratio a/c that closest to the atom located in the center of the cell are the atoms located at its vertices. The fourth-order axis c lies in the plane of the section and is the easy axis.

We introduce a Cartesian orthogonal coordinate system with an axis x directed along the light axis and an axis perpendicular to the section plane.

The power of the single-ion anisotropy takes the form

$$W_{\text{an}} = K_{\perp} \sum_{i \in \text{f}} (s_i^{(z)})^2 - K_{||} \sum_{i} (s_i^{(x)})^2, \tag{4.1}$$

where the summation occurs over localized spins, $K_{||} > 0$ is a constant of uniaxial anisotropy, and $K_{\perp} > 0$ is a constant of anisotropy of the ferromagnet shape, introduced to account for the power disadvantage of states in which the magnetization has z -component perpendicular to the plane of the layers. Immediate consideration of the dipole–dipole interaction of spins leads to a cardinal complication of the problem and dramatically increases the calculation time.

A. S. Sigov, *Multilayer Magnetic Nanostructures*, Springer Aerospace Technology,
https://doi.org/10.1007/978-981-19-6246-2_4

We assumed that the constant $K_{||}$ is the same in the ferro- and antiferromagnetic layers to reduce the number of parameters.

In the case of a slice (100) of a volume-centered cubic lattice, there are two mutually perpendicular easy axes in the plane of the slice, and the following expression should replace the second term in the right part (4.1)

$$-K_{||}\sum_{i}\left[(s_i^{(x)})^4+(s_i^{(y)})^4\right]. \tag{4.2}$$

The equilibrium distribution of spins, taking into account the power of their exchange interaction (2.2) and the power of single-ion anisotropy, was determined by mathematical modeling of their dynamics based on the system of Landau-Lifshitz-Gilbert equations

$$\hbar\frac{d}{dt}\vec{S}_i=\left[\vec{S}_i,\vec{H}_{eff}\right]+\mu\vec{H}_{eff}, \tag{4.3}$$

where μ is the attenuation, and the effective field H_{eff} is

$$H_{\text{eff}}^{p}=-\frac{\partial W}{\partial s_i^p}, \tag{4.4}$$

where $p = x, y, z$, and W is the total power of exchange and anisotropy.

The solution of the equations (4.3) was found by the Runge–Kutta method of the fourth order. The achievement of equilibrium was determined by the temporal behavior of the system's total power, namely by its approaching to a constant value.

In the case of an uncompensated cross-section of the antiferromagnet surface, the edges of the atomic steps at the interface of the layers were assumed to be parallel to each other and the axis y of the coordinate system. A two-dimensional problem on the distribution of spins in a plane xz with periodic x boundary conditions was solved.

4.2 Solitary Domain Wall

It should be noted that in the absence of a magnetic field, this problem is entirely equivalent to the issue of a thin film of an antiferromagnet on a ferromagnetic substrate. As shown above, the cases of compensated and uncompensated slices of the antiferromagnet surface differ radically. Therefore, we will consider them sequentially [2].

Let us start with the issue when the parameter R is much larger than the thickness of the ferromagnetic film a. In this limiting case, to minimize the power of the interlayer interaction at the entire interface, it is advantageous for the ferromagnetic film to break into domains with domain walls perpendicular to the interface of the

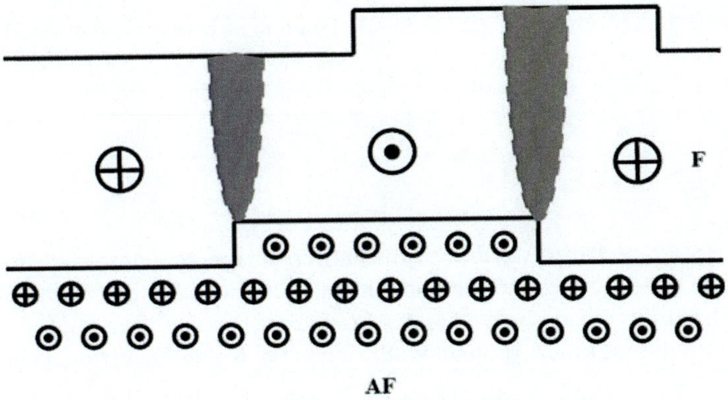

Fig. 4.1 The nanodomain phase in a two-layer system in the case of $\gamma_{\mathrm{af}} >> 1$

layers. The domain walls pass through the edges of the atomic steps at the interface (Fig. 4.1).

Indeed, the stages divide the interface into two types of areas: in the first type of areas, the parallel orientation of the ferro- and antiferromagnetic order parameters is energetically advantageous, and in the second type of areas, the antiparallel direction of these order parameters is beneficial. After splitting the ferromagnetic layer into 180° domains, the optimal orientation of the order parameters is realized in each area.

Of course, in this case, the system's power increases by the amount of the power of the resulting domain walls, but due to the inequality $R >> a$, the polydomain phase is more advantageous. The idea of splitting a frustrated system into domains was expressed earlier [A. Berger, H. Hopster. *Phys. Rev. Lett.* **73**, 193 (1994); E.J. Escorcia-Aparicio et al. *Phys. Rev. Lett.* **81**, 2144 (1998)], however, the structure of emerging domain walls generated by frustrations was not discussed in these works.

It turned out that the structure of the domain wall significantly depends on the ratios of the energies of the exchange interaction of spins in the layers and between the layers, which are given by two dimensionless parameters

$$\gamma_{\mathrm{af}} = |J_{\mathrm{af}}|/J_{\mathrm{f}}, \tag{4.5}$$

and

$$\gamma_{\mathrm{f,af}} = \left|J_{\mathrm{f,af}}\right|/J_{\mathrm{f}}. \tag{4.6}$$

The first term on the right side of Eq. (4.1) makes the states in which the order parameters have Z component unprofitable. Therefore, the spin vectors lie in the plane of the layers, and the direction of the order parameters can be set by the angle $\theta_{f(\mathrm{af})}$ that the corresponding order parameter forms with the easy axis in the plane of the layer. The modulus of the order parameter, as already mentioned, remains unchanged.

4.2.1 $\gamma_{af} >> 1$

In the case $\gamma_{\rm af} >> 1$ of distortion of the antiferromagnetic parameter, the order in the substrate is negligible, and the domain wall generated by the frustration crosses only the film. In it, the angle θ_f changes from zero to π.

Compared to traditional domain walls, it has the following differences:

- its width $\delta(z)$ is determined not by the competition of the exchange power and the anisotropy power, as in the case of traditional domain walls, but by the competition of the intra-layer and interlayer exchange interactions;
- the width of the domain wall δ increases as it moves away from the edge of the atomic step at the interface.
- An estimate for the dimensionless width of the domain wall $\delta_0 = \delta_f(z = 0)$ and a particular thickness-averaged value $\left(\delta_f\right)_z^{'}$, denoted later ξ, can be obtained from simple power considerations. We approximate the value $\theta(x, z)$ as follows:

$$\theta(x,z) = \begin{cases} \pi \text{ at } x \geq \delta_{\rm f}(z), \\ \frac{\pi}{2}(1 + x/\delta_{\rm f}(z)) \text{ at } -\delta_{\rm f}(z) < x < \delta_{\rm f}(z), \\ 0 \text{ at } x \leq -\delta_{\rm f}(z), \end{cases} \tag{4.7}$$

where

$$\delta_{\rm f}(z) = \delta_0 + \varsigma z, 0 \leq z \leq a. \tag{4.8}$$

The contribution to the power due to the inhomogeneity of the order parameter in the domain wall, calculated for 1 m of its length along the y-axis, is

$$w_1 = \frac{J_f}{2b} \int_0^a {\rm d}z \int_{-\infty}^{\infty} {\rm d}x \left[\left(\theta_x^{'}\right)^2 + \left(\theta_z^{'}\right)^2\right] \sim \frac{\pi^2 J_f}{4b}\left(\frac{1}{\varsigma} + \frac{\varsigma}{3}\right) \ln\frac{\varsigma a + \delta_0}{\delta_0}. \tag{4.9}$$

The presence of a step leads to an increase in the interaction power of the film with the substrate by an amount

$$w_2 = \frac{2J_{\rm f,af}}{b} \int_0^{\infty} {\rm d}x[1 - \cos\theta(x, 0)] \sim \frac{2J_{\rm f,af}}{b}\delta_0. \tag{4.10}$$

Minimizing the power w_1 in the parameter ς, and then, the total power of the domain wall $w = w_1 + w_2$ in the parameter δ_0, we find these values.

If $\gamma_{\rm f,af}a << 1$, $\varsigma \sim \sqrt{\gamma_{\rm f,af}a} << 1$, then the expansion of the domain wall can be neglected $\delta(z) \approx \delta_0 \sim \sqrt{a/\gamma_{\rm f,af}}$. Here and further on, the distances (in this case, the

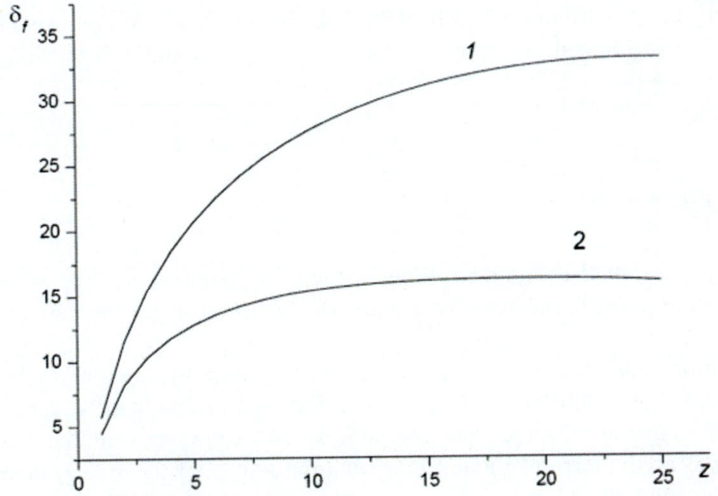

Fig. 4.2 The dependence of the domain wall thickness on the distance to the interface is calculated for (1) $\Delta_f > a$ and (2) $\Delta_f < a$. All spaces are given in units of the corresponding interatomic area

thickness of the ferromagnetic layer a) are dimensioned by the corresponding lattice constant.

In the more relevant case $\gamma_{\mathrm{f,af}} a >> 1$, we get $\varsigma \sim 1$ and $\delta_0 \sim \max\left(1, \gamma_{\mathrm{f,af}}^{-1}\right)$. The calculated dependence $\delta(z)$ in the exchange approximation is shown in Fig. 4.2 (Curve 1). It can be seen that $\delta(z)$ grows linearly near the interface of the layers and has a zero derivative on the free surface of the film. The characteristic thickness of the domain wall generated by the frustration is of the order of the film thickness a. In nanometer-scale films of consistency, the domain walls caused by the frustrations are substantially narrower than the traditional domain walls.

For the power of the domain wall per unit length, we obtain the following estimate [3]:

$$W \sim \begin{cases} \frac{J_f}{b}\sqrt{\gamma_{\mathrm{f,af}} a} & y_{\mathrm{f,af}}\, a \ll 1 \\ \frac{J_f}{b}\ln\left(\gamma_{\mathrm{f,af}} a\right) & y_{\mathrm{f,af}}\, a \gg 1 \end{cases}, \tag{4.11}$$

The broadening of the domain wall in the case $\gamma_{\mathrm{f,af}} a >> 1$ leads to the fact that its power w per unit length of the atomic step increases with increasing film thickness only logarithmically (b is the interatomic distance).

The exchange approximation is valid for a range of thicknesses $a << \Delta_{\mathrm{f}}$, where

$$\Delta_{\mathrm{f(af)}} \sim \sqrt{J_{\mathrm{f(af)}}/K_{\parallel}} \tag{4.12}$$

there is a width of the traditional domain wall in a ferromagnetic film (antiferromagnetic substrate). With a further increase in the film thickness, the power of the single-ion anisotropy should be taken into account. That will lead to the fact that the

$a \sim \Delta_f$ further growth in the width of the domain wall generated by frustration will stop with an increase in the film thickness $\delta_f(z)z > \Delta_f$ and will reach a constant value $\delta_f \approx \Delta_f$ (Fig. 4.2, curve 2).

4.2.2 $\gamma_{af} << 1$

If $\gamma_{f,af} a/\gamma_{af}^2 << 1$, then the distortions of the order parameter in the substrate are slight and the domain wall parameters are similar to those for the case $\gamma_{af} >> 1$, $\gamma_{f,af} a << 1$.

In the more realistic case $\gamma_{f,af} a/\gamma_{af}^2 >> 1$, the domain wall structure is significantly more complex, and substantial distortions of the antiferromagnetic order parameter occur in the substrate. Two parameters characterize the domain wall: the width of the domain wall in the ferromagnetic film δ_f and the size δ_0 of the area at the interface of the layers near the edge of the atomic step, in which the mutual orientation of the magnetic order parameters differs from the optimal one (parallel on one side of the step and antiparallel on the other).

A simple estimation of these parameters from power considerations gives

$$\delta_f \sim a/\gamma_{af}, \tag{4.13}$$

and

$$\delta_0 \sim \min\left(1, \gamma_{af}/\gamma_{f,af}\right). \tag{4.14}$$

Since $\delta_f >> a$, the broadening of the domain wall in the ferromagnetic layer can be neglected. Distortions in the antiferromagnetic substrate occupy an area with a radius of about an order of magnitude δ_f around the edge of the atomic step (Fig. 4.3).

It is the distortions in the substrate that make the main contribution to the power of the domain wall, which, in the exchange approximation, is the value per unit length of the atomic step

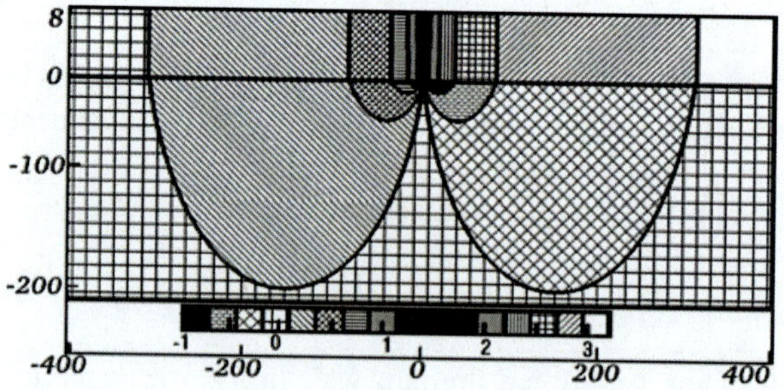

Fig. 4.3 Distribution of order parameters in the domain wall. The ordinate equal to zero corresponds to the interface. All distances are given in units of the corresponding interatomic space. The box shows the correspondence between the hatching and the value $\theta_{f(af)}$ measured in radians

$$w \sim \frac{J_{\mathrm{af}}}{b} \ln \frac{\delta_{\mathrm{f}}}{\delta_0}. \tag{4.15}$$

If the thickness of the substrate $d_{\mathrm{af}} < \delta_{\mathrm{f}}$, then the domain wall will penetrate through it; that is, the antiferromagnetic layer will be divided into domains, and the ferromagnetic layer will remain almost uniform. In other words, we come to the case $\gamma_{\mathrm{af}} >> 1$, but the layers are reversed. Thus, we have found a critical thickness above which the substrate can be considered thick.

The exchange approximation is valid if $\delta_f << \Delta_{\mathrm{f}}, \Delta_{\mathrm{af}}$. Since we assumed that the parameter $K_{\|}$ is the same in both layers, then $\Delta_{\mathrm{af}} \sim \gamma_{\mathrm{af}}^{1/2} \Delta_{\mathrm{f}} << \Delta_{\mathrm{f}}$. Therefore, with growth a, the value δ_f will initially exceed Δ_{af}.

Consider the structure of the domain wall in the region $\Delta_{\mathrm{af}} << \delta_{\mathrm{f}} << \Delta_{\mathrm{f}}$. In this case, the area of the distortions of the antiferromagnetic order parameter in the substrate has not semicircular but flattened shape. At a distance of an order of magnitude Δ_{af} from the edge of the atomic step, the distortions take the form of an exchange spiral parallel to the interface. Its structure is entirely similar to the structure of the Bloch domain wall in this material. We use the term "exchange helix" rather than "domain wall", since the domain wall separates two domains in the same layer, and the exchange helix occurs at the interface of different materials.

Let us now estimate the value δ_{f} the width of the domain wall in a ferromagnet. The distortion power of the ferromagnetic order parameter per unit length of the atomic stage is the value w_{f}.

$$w_{\mathrm{f}} \sim \frac{J_{\mathrm{f}} a}{b \delta_{\mathrm{f}}}. \tag{4.16}$$

The distortion power of the antiferromagnetic order parameter w_{af} has the form [4]

$$w_{\mathrm{af}} \sim \frac{|J_{\mathrm{af}}|}{b} \ln \frac{\Delta_{\mathrm{af}}}{\delta_0} + \varepsilon_{\mathrm{af}} (\delta_{\mathrm{f}} - \Delta_{\mathrm{af}}) b, \tag{4.17}$$

where

$$\varepsilon_{\mathrm{af}} \sim \sqrt{|J_{\mathrm{af}}| K_{||}} / b^2 \tag{4.18}$$

there is surface power of the traditional domain wall (and hence the exchange helix) in an antiferromagnet.

Minimizing the total power $w_{\mathrm{f}} + w_{\mathrm{af}}$ by the parameter δ_{f}, we find

$$\delta_{\mathrm{f}} \sim \left(\frac{J_{\mathrm{f}} a}{\varepsilon_{\mathrm{af}} b^2} \right)^{1/2} \sim \left(\frac{a \Delta_{\mathrm{af}}}{\gamma_{\mathrm{af}}} \right)^{1/2} \propto a^{1/2}. \tag{4.19}$$

The value δ_f is compared with the thickness of the traditional domain wall Δ_f at $a = \Delta_{af}$. Further growth δ_f with an increase in the thickness of the ferromagnetic layer stops and one has $\delta_f \approx \Delta_f$ at $a > \Delta_{af}$.

4.3 Spin Vortices

The picture of the polydomain state of a ferromagnet considered in the previous section is valid if the distance R between the atomic steps at the interface of the layers exceeds the critical value.

In the region of applicability of the exchange approximation, when $\delta_f << \Delta_f$, this critical value R_c is the width of the domain wall δ_f. Since in the case $a \sim 1$ of nm, the value δ_f is also several nanometers, at $R \sim 10$ nm, a nanodomain state is realized in the film, which can be identified experimentally only with the appropriate spatial resolution measuring device.

In the opposite extreme case $R << \delta_f$, the domain walls overlap. As a result, the order parameter will remain almost uniform in the main volume of the ferromagnet. At $\gamma_{af} >> 1, \gamma_{f,af}a << 1$, or $\gamma_{af} << 1, \gamma_{f,af}a/\gamma_{af}^2 << 1$, or $R << \delta_0$, the distortions of the order parameters are slight both in the film and in the substrate.

If $\gamma_{af} >> 1$, $\gamma_{f,af}a >> 1$, $\delta_0 << R << a$, then a kind of static spin vortices arise in the film near the substrate (Fig. 4.4a), penetrating deep into the film at a distance of the order *of R*. Each vortex is bounded by the edges of the steps and expands as it moves away from them. In the rest of the film, the uniform distribution of the order parameter remains undisturbed.

In the case $\gamma_{af} \gg 1$, $\gamma_{f,af}a/\gamma_{af}^2 \gg 1$, $\delta_0 << R << \delta_f$, the film remains homogeneous, and similar spin vortices form near the interface in the substrate. In the field of applicability of the exchange approximation, the size of the vortex in the direction perpendicular to the boundary is equal to the minimum size of the area bounded by steps on the plane of the layer interface.

In equal exchange energies ($\gamma_{af} = 1$), the vortex is located symmetrically relative to the boundary of the layers (Fig. 4.4b).

Spin vortices, as well as the areas at the interface, belong to two types. The type of vortex corresponds to the kind of area in which it occurs. The rotation of the order parameter in the vortex occurs in a plane parallel to the interface and differs in eddies of different types.

Let us now consider the mutual orientation of the ferromagnetic and antiferromagnetic order parameters in the vortex phase. As already noted, the steps divide the entire interface into two regions with a total area σ_1 and σ_2, respectively. If the average magnetization of a ferromagnetic film makes an angle ψ with the antiferromagnetic order parameter in the depth of the substrate, then in the vortex occupying the region of the first type the angle $\theta_f - \theta_{af}$ changes from zero to ψ, while in the vortices in the region of the second type the angle $\theta_f - \theta_{af}$ changes from ψ to π.

By analogy with the "magnetic proximity" model of Slonczewski [J. C. Slonczewski *Phys. Rev. Lett.* **67**, 3172 (1991)], the total power of the vortices can be

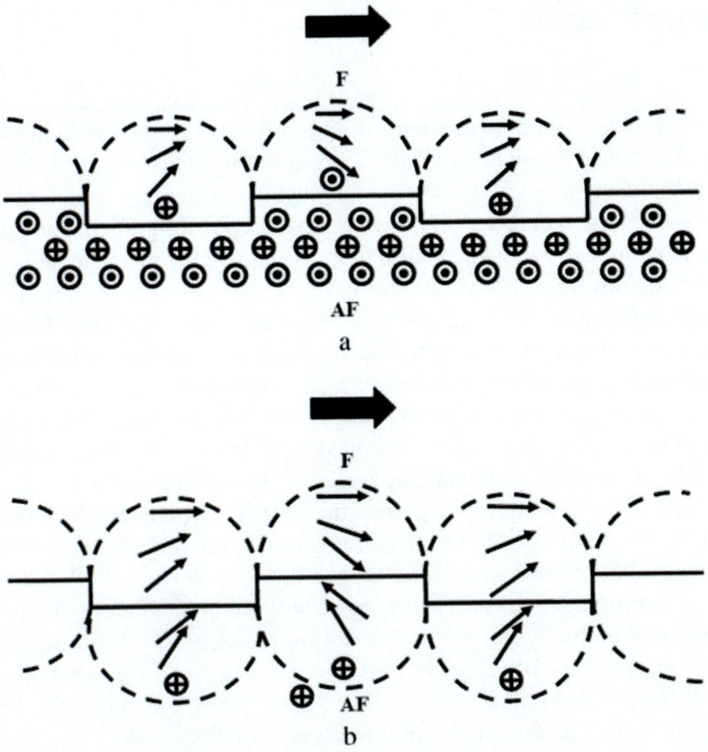

Fig. 4.4 The orientation of the spins of the ferromagnetic and antiferromagnetic layers near the interface in the monodomain phase in the case $|J_{af}| >> J_f$ of (**a**); the exposure of the ferromagnetic and antiferromagnetic order parameters near the interface in the monodomain phase in the case $|J_{af}| \approx J_f$ of (**b**). The arrows indicate the direction of the corresponding order parameter. The rotation of the order parameters occurs in the plane of the layers

written as.

$$W = C_1\psi^2 + C_2(\pi - \psi)^2, \tag{4.20}$$

where the constant C_i is proportional to the area σ_i occupied at the interface by the region of the corresponding type [5]

$$C_i = \frac{\min(J_f, |J_{af}|)\sigma_i}{Rb} \ln \frac{R}{b}. \tag{4.21}$$

If both types of the regions are represented with the same probability, then the power minimum corresponds to the mutually perpendicular orientation of the order parameters away from the interface ($\psi = \pi/2$).

4.4 Phase Diagram

In the field of applicability of the exchange approximation, the transition from the polydomain to the monodomain state of a ferromagnetic film with an increase in the thickness of the ferromagnet or an increase in the roughness (a decrease in the parameter R) occurs continuously. It is not, strictly speaking, a phase transition. It can be carried out in the process of layer-by-layer deposition of a ferromagnet on an antiferromagnetic substrate. As the film is sprayed, the change from the polydomain to the monodomain state will be observed.

One can get out of the applicability of the exchange approximation for describing the monodomain phase by increasing the film thickness at a constant parameter R. When the thickness of the film becomes on the order of or exceeds the width of the domain wall in a layer with a lower exchange stiffness, it becomes necessary to take into account the power of the single-ion anisotropy. In this case, the configuration of the easy axes in the plane of the layer plays a fundamental role.

In the presence of two mutually perpendicular easy axes in the plane of the layer, the situation practically does not change compared to the one obtained in the framework of the exchange approximation. Away from the interface, both the ferromagnetic and antiferromagnetic order parameters will be directed along their easy axes while remaining mutually perpendicular. In the case $\gamma_{\mathrm{af}} >> 1$, an increase in the parameter R in the range of values $R \sim \Delta_f << a$, leads to a change in the shape of the vortices from semicircular to oblate. At $\Delta_f << R << a$, they take the form of 90° exchange spirals parallel to the interface (Fig. 4.5a).

If there is only one easy axis in the plane of the layer [the tetragonal cut (100)], then the mutually perpendicular orientation of the order parameters does not satisfy the minimum of the anisotropy power. The competition between the power of the interlayer interaction and the power of anisotropy leads to the appearance of the 90° exchange spiral near the interface of the layers. It is located mainly in a layer with a lower power of the traditional domain wall. The characteristic size of the region occupied by the exchange spiral is of the order of the width of the Bloch domain wall in the same material and far exceeds the size of the vortices. Near the interface (but outside the region of static spin vortices), the order parameters are oriented mutually perpendicular, so both of these vectors cannot be collinear to the easy axis. In the exchange spiral, a reversal of the order parameter (parameters) occurs, as a result of which they turn out to be collinear with the easy axis (Fig. 4.5b).

If the film thickness is $a << \min(\Delta_{\mathrm{f}}, \Delta_{\mathrm{af}})$, then the magnetization of the film is oriented along the hard direction. With growth a, an exchange spiral is gradually formed in a layer with a lower exchange stiffness; that is, the transition to the collinear orientation of the order parameters occurs continuously.

The phase diagrams for the cases $\gamma_{\mathrm{af}} >> 1$ and $\gamma_{\mathrm{af}} << 1$ and one easy axis in the layer plane are shown in Fig. 4.6.

If in the case $\gamma_{af} >> 1$ and one easy axis in the plane of the layer, being in phase with vortices and an exchange spiral in the ferromagnet (phase 3 in Fig. 4.6a), increases the distance between the edges of atomic steps R, then at $R \sim \Delta_{\mathrm{f}}$, in the

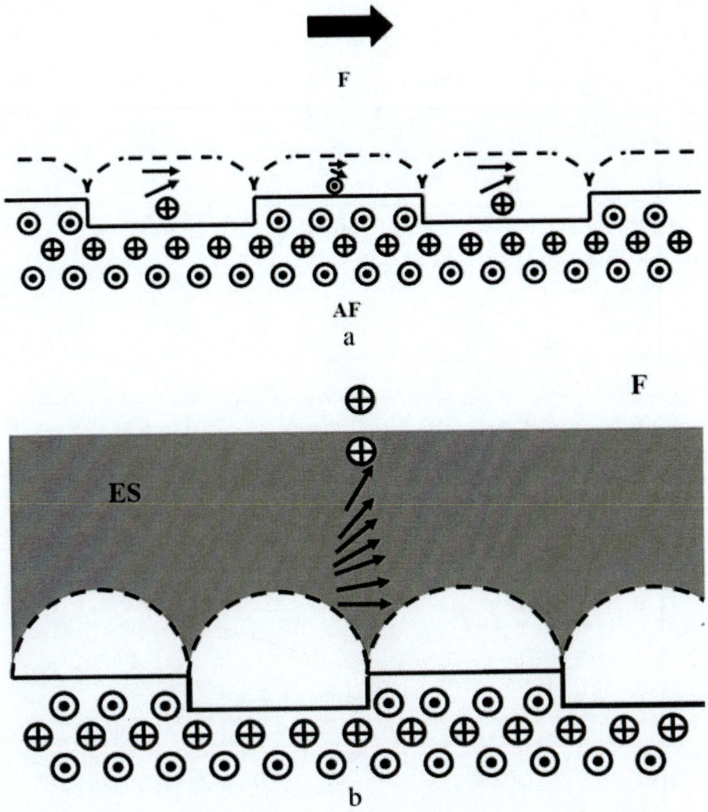

Fig. 4.5 Exchange spirals in the case of two mutually perpendicular easy axes (**a**), exchange spiral (ES) in a ferromagnet (**b**). The arrows indicate the direction of the corresponding order parameter. The rotation of the order parameters occurs in the plane of the layers

regions of the first type, an "annihilation" will occur of growing with the parameter R static spin vortices and the exchange spiral: distortion of the order parameter in the vortex compensate for those in the exchange spiral. In the regions of the second type these distortions are "added together". As a result, in these regions, at values of R greater than Δ_f but still less than the film thickness α, a 180° exchange spiral is formed, while in the of the first type, distortions of the order parameter disappear (Fig. 4.7).

Let us now turn to the case $\gamma_{af} << 1$. In the region of the parameter R values determined by the inequality $a << \Delta_{af} << R << \delta_f << \Delta_f$, the shape of the vortices changes from semicircular (at $R << \Delta_{af}$) to flattened, they take the form of a 90° exchange spiral (phase 5 in Fig. 4.6b). This transition with growing R occurs continuously at $R \sim \Delta_{af}$.

If there are two mutually perpendicular easy axes in the plane of the layers a, then the situation does not change with the increase in the thickness of the ferromagnet.

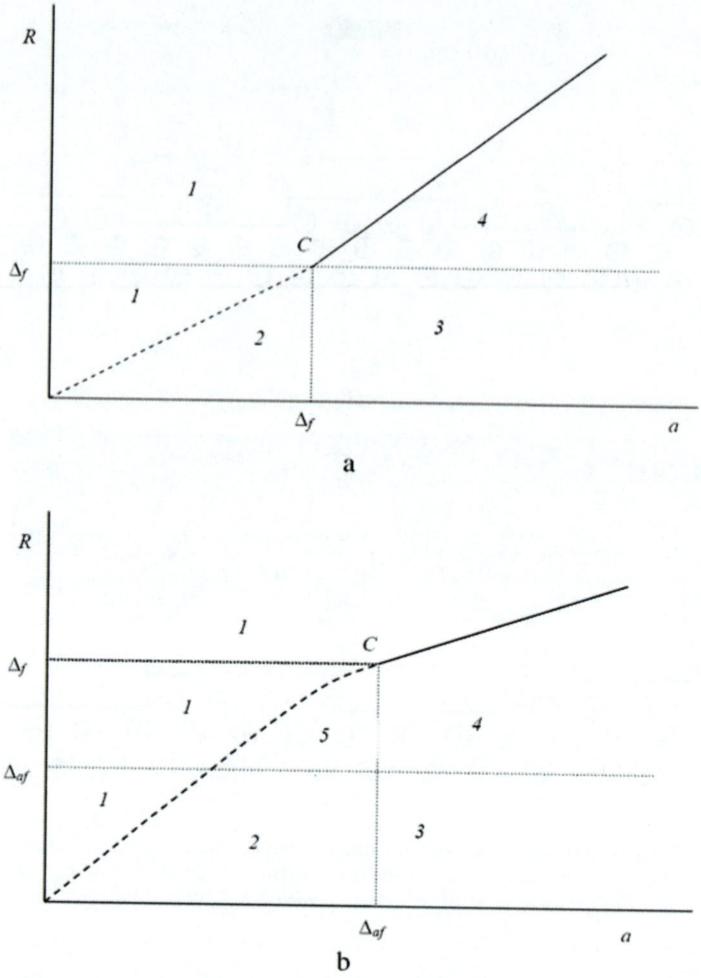

Fig. 4.6 The phase diagram in the cases $\gamma_{af} >> 1$ (**a**) and $\gamma_{af} << 1$ (**b**): 1—a polydomain phase, 2—a phase with semicircular vortices and magnetization of the filmoriented perpendicular to the easy axis, 3—a phase with semicircular vortices and 90° exchange spiral, 4—a phase with 180° exchange spirals, parallel to the interface, in the regions of the same type, and 5—a phase with flattened vortices and magnetization of the filmoriented perpendicular to the easy axis. The solid line corresponds to the line of phase transitions of the first order, and C is the critical point

In the presence of one easy axis in the plane of the layer, with the growth of the parameter a at $a \sim \Delta_{af}$, an exchange spiral is formed and its "annihilation" with vortices occurs in the regions of the first type and "addition" in the regions of the second type. As a result, as for $\gamma_{af} >> 1$, in the regions of the first type, the order parameter at $>> \Delta_{af}$ remains uniform, and in the regions of the second type, a 180° exchange spiral appears (phase 4 in Fig. 4.6b). The only difference from the case

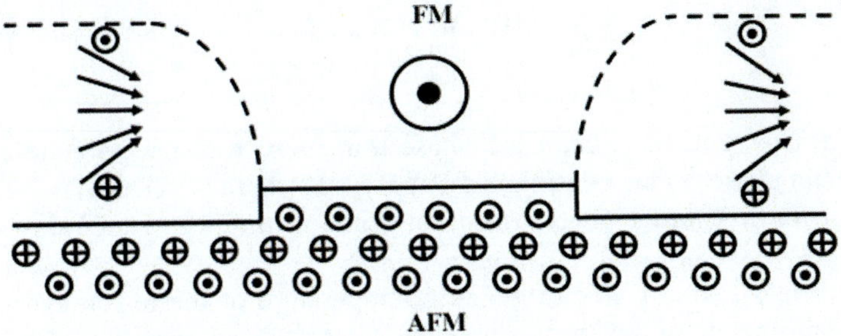

Fig. 4.7 Exchange spirals in the case of a single easy axis. The arrows indicate the direction of the corresponding order parameter. The rotation of the order parameters occurs in the plane of the layers

$\gamma_{\text{af}} >> 1$ is that the exchange spiral is located in a softer substrate, and not in the film [2].

It should be noted that the first attempts to represent qualitatively the distribution of spins in a frustrated system [D Stoeffler, F. J. Gautier. *Magn. Magn. Mater.* **147**, 260 (1995); E. E. Fullerton, C. H. Sowers, S. D. Bader. *Phys. Rev. B* **56**, 5468 (1997)] refer specifically to the parameter range corresponding to the phase 4 in Fig. 4.6. The authors of these studies assumed a domain wall in the layer with a lower exchange stiffness, closing the edges of the steps. Moreover, the domain wall was assumed to be extremely narrow - atomically sharp. Computer simulations were carried out within the framework of the strong coupling model, and the assumption of collinear spin orientation confirmed this prediction [A. Berger, E. E. Fullerton. *J. Magn. Magn. Mater.* **165**, 471 (1997)]. However, the assumption of the collinear orientation of the spins is valid if the power of the single-ion anisotropy is much higher than the power of the exchange interaction. We investigate the opposite limiting case, in which the spin configuration turns out to be essentially non-collinear.

What happens if, while in $a >> \min(\Delta_{\text{f}}, \Delta_{\text{af}})$ the phase with exchange spirals, we increase the parameter R? There will be a phase transition of a monodomain phase with exchange spirals near the interface to a polydomain phase with domain walls perpendicular to the interface. Since at the critical value $R = R_c$ corresponding to the phase transition, the domain walls do not overlap, this transition occurs abruptly; that is, it is the first order phase transition, blurred due to the spread over the distances between the edges of the atomic steps.

To find the value R_C, we equate the energies of the polydomain and monodomain phases. The domain wall perpendicular to the film has a power per unit length of the edge of the atomic step $\varepsilon_{\text{f}} ab$, where ε_{f} is given by the formula (4.18) replaced $|J_{\text{af}}|$ by J_{f}. Since the distance between the walls equals R, the power of the polydomain phase per unit area of the interface will be $\varepsilon_{\text{f}} a/R$.

At the same time, the power of the monodomain phase per unit area is by the order of magnitude ε_{f} at $\gamma_{\text{af}} >> 1$ and ε_{af} at $\gamma_{\text{af}} << 1$. Equating these energies, we find

$$R_c \sim \begin{cases} a, & \gamma_{\mathrm{af}} >> 1 \\ a/\gamma_{\mathrm{af}}^{1/2}, & \gamma_{\mathrm{af}} << 1 \end{cases}. \tag{4.22}$$

Since at $a < \min(\Delta_f, \Delta_{af})$ the phase transition from the polydomain to the monodomain phase occurs continuously, at $R \sim \Delta_f$ there is a critical point in which the line of the first order phase transitions ends in the phase diagrams shown in Fig. 4.6.

In the case $\gamma_{\mathrm{af}} << 1$ when the Curie temperature of the ferromagnet exceeds the Neel temperature of the antiferromagnet, near the Neel temperature $\delta_{\mathrm{f}} \propto \gamma_{\mathrm{af}}^{-1} \propto (T_N - T)^{-1}$ and increases with increasing temperature. Therefore, the transition from a polydomain to a monodomain state is possible during the heating process.

Since the magnetic phase diagram is determined by the characteristic distance R between the edges of the atomic steps, replacing the set of parallel steps with their square grid with the same cell size R does not lead to a qualitative change in the resulting phase diagram.

4.5 Experimental Results

Predicted in 1998 in [V. D. Levchenko et al. *JETF* **114**, 1817 (1998)], domain walls of a new type generated by frustrations were discovered experimentally in 2004 during the study of thin antiferromagnetic films of manganese on an iron substrate by spin-polarization scanning tunneling microscopy [U. Schlickum et al. *Phys. Rev. Lett.* **92**, 107,203 (2004); W. Wulfhekel, U Schlickum, J Kirschner. *Microscopy Research Technique* **66**, 105 (2005)]. As already noted, the problem of a thin film of an antiferromagnet on a ferromagnetic substrate is completely equivalent to the one under consideration. The detected domain walls appeared exactly at the location of the atomic steps, and their thickness grew in proportion to the film thickness with a proportionality coefficient of the order of one [4].

In the paper [E. J. Escorcia-Aparicio et al. *Phys. Rev. B* **59**, 11,892 (1999)], the phase diagram "thickness—vicinal angle" of an iron film on a chromium substrate with a surface orientation close to (001) was studied. This choice of materials is appropriate for the occasion $\gamma_{\mathrm{af}} << 1$. At a vicinal angle of zero, the polydomain phase was observed at film thicknesses $a < a_c = 3.5$ nm. In a film with a critical thickness a_c, the characteristic distance R between the edges of randomly placed steps corresponds to the width of the domain wall. At large values *of a*, a monodomain phase with the orientation of the magnetization perpendicular to the edges of the steps was observed. With an increase in the vicinal angle, the concentration of atomic steps at the interface increased, which was accompanied by a drop in the critical value a_c.

References

1. Levchenko VD, Sigov YS, Morosov AI, Sigov AS (1998) "Unusual" domain walls in multilayer ferromagnet - layered antiferromagnet. ZhETF 114(5)(11): 1817–1826
2. Morosov AI, Sigov AS (1999) A new type of domain boundaries in multilayer magnetic structures. UFN 169(8):922–924
3. Levchenko VD, Morosov AI, Sigov AS (2000) Phase diagram of a thin ferromagnetic film on the surface of an antiferromagnet. Lett JETF 71(9):544–549
4. Morosov AI (2008) Magnetic phase diagram of a two-layer ferromagnet-antiferromagnet system with a rough interface. FTT 50(4):675–680
5. Morosov AI (2009) Magnetic phase diagram of the frustrated two–layer system ferromagnet–antiferromagnet. J Magn Magn Mater 321(7):868–871

Chapter 5
Compensated Slice

5.1 Spin-flop Orientation

Koon [N] proposed the optimal orientation of the spins at a smooth interface in the case of a compensated cross-section of an antiferromagnet [N. C. Koon. *Phys. Rev. Lett.* **78,** 4865 (1997)]. In the framework of the exchange approximation, the ferromagnetic order parameter is oriented perpendicular to the antiferromagnetism vector. The magnetizations of the sublattices of the upper atomic plane of the antiferromagnet due to the exchange interaction with the ferromagnet are skewed so that the resulting magnetization of the plane is parallel (in the case $J_{f,\mathrm{af}} > 0$) to the magnetization of the ferromagnet—a spin-flop orientation similar to the bevel of the sublattices of the antiferromagnet in the magnetic field (Fig. 1.4). Since the exchange field of the ferromagnet acts only on the upper atomic plane of the antiferromagnet, the value of the angle of the slant of the sublattices θ decreases deep into the antiferromagnet at atomic distances, and the direction of the slope of the magnetization is opposite in the neighboring atomic planes of the antiferromagnet parallel to the interface [1].

At the same time, in a ferromagnet, due to the exchange interaction of its spins with the spins of the antiferromagnet, the spins are rotated by an angle φ, and an induced vector of antiferromagnetism appears (Fig. 5.1). Its value decreases deep into the ferromagnet at atomic distances, and the characteristic length of this exponential decrease is determined by the number of nearest neighboring spins lying in the same plane as the selected spin z_1, and the number of most relative neighboring spins in the following atomic plane z_2 ($z = z_1 + 2z_2$—the total number of nearest neighbors). In turn, the values of z_1 and z_2 depend on the type of crystal lattice and the surface slice (see Table 5.1).

In the area of small values $\varphi_j\left(\left|\varphi_j \ll 1\right|\right)$, where j is the number of the atomic plane of the ferromagnet, counted from the interface, the relation takes place

$$\varphi_j = \kappa \varphi_{j-1} \tag{5.1}$$

A. S. Sigov, *Multilayer Magnetic Nanostructures*, Springer Aerospace Technology,
https://doi.org/10.1007/978-981-19-6246-2_5

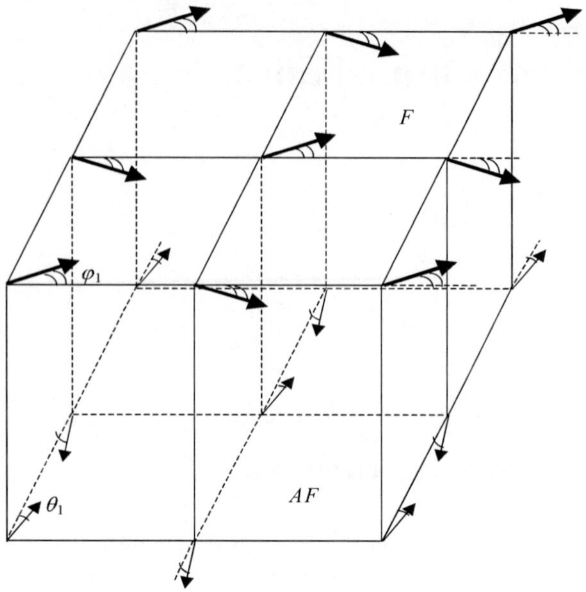

Fig. 5.1 Orientation of the spins of the lower atomic plane of the ferromagnet and the upper atomic plane of the antiferromagnet for the slice (100) of a simple cubic lattice

Table 5.1 Parameters that characterize different slices of crystals

Crystal lattice type	Slice	z_1	z_2	k
Simple cubic	(1,0,0)	4	1	$5 - \sqrt{24} = 0.101021$
Simple cubic	(1,1,0)	2	2	$2 - \sqrt{3} = 0.267949$
Volume-centered or Volume-centered tetragonal	(1,1,0)	4	2	$3 - \sqrt{8} = 0.171573$

where

$$\kappa = \frac{z_1}{z_2} + 1 - \sqrt{\frac{z_1}{z_2}\left(\frac{z_1}{z_2} + 2\right)}. \tag{5.2}$$

Similarly, for the bevel angles of an antiferromagnet θ_j, we have

$$\theta_j = -\kappa\theta_{j-1} \tag{5.3}$$

with the same value κ.

It should be noted that just the induced "improper" order parameters in the layers lead to the interaction of the layers in the case of a compensated slice of the antiferromagnet surface.

How will this picture change when going beyond the exchange approximation? If there are two mutually perpendicular light axes in the plane of the layer, it will remain almost unchanged.

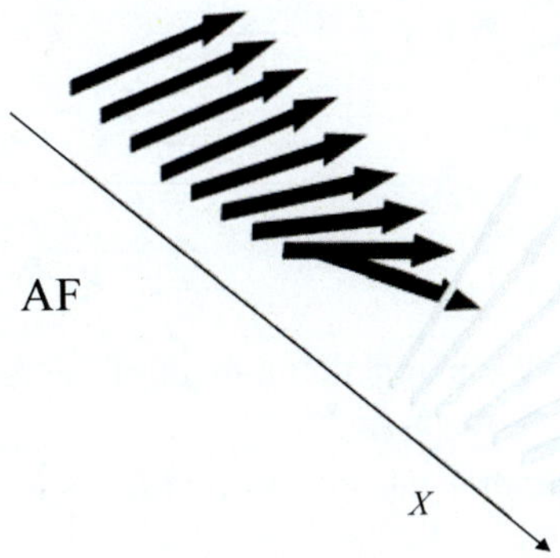

Fig. 5.2 Distribution of order parameters near the ferromagnet–antiferromagnet interface

In the case of a single easy axis in the layer plane, the mutually perpendicular orientation of the order parameters away from the interface does not correspond to the minimum energy of the single-ion anisotropy. As a result, in a sufficiently thick ferromagnetic film $\alpha \gg \min(\Delta_f, \Delta_{\mathrm{af}})$, an exchange spiral occurs near the interface, similar to the one discussed in Sect. 4.4.

If $\alpha \ll \min(\Delta_f, \Delta_{\mathrm{af}})$, then the exchange approximation remains valid.

Consider the structure of the exchange spiral in the case of $\alpha \gg \min(\Delta_f, \Delta_{\mathrm{af}})$. Near the interface, the magnetization vector of the ferromagnet and the antiferromagnetism vector of the substrate deviate from the easy axis in different directions so that the angle between them is 90°. As they move away from the interface, they unfold to become parallel to the easy axis (Fig. 5.2). The characteristic scale of the length at which the reversal occurs coincides with the width of the traditional domain wall in this layer. Note that the exchange spiral occurs in the absence of an external field and corresponds to the equilibrium state of the ferromagnet–antiferromagnet system.

The energy of the spiral can be found as the difference between the energies of the layers and the energies of their homogeneous ground state. In the continuum approximation, when $\min(\Delta_{\mathrm{af}}, \Delta_f) \gg 1$, the energy of the spiral has the form

$$\tilde{W} = NJ_f z_2 \left\{ \begin{array}{l} \gamma_{\mathrm{af}} \int_{-\infty}^{0} \left[\left(\theta'_{\mathrm{af}}(x)\right)^2 + \alpha_{\mathrm{af}}(1 - \cos 2\theta_{\mathrm{af}}(x)) \right] dx \\ + \int_{0}^{\infty} \left[\begin{array}{l} \left(\theta'_f(x)\right)^2 + \alpha_f\left(1 - \cos 2\theta_f(x)\right) \\ +2\beta\left(1 - \cos\theta_f(x)\right) \end{array} \right] dx \end{array} \right\}, \tag{5.4}$$

where the ferromagnet occupies a half $x > 0$-space, the antiferromagnet occupies a half-space $x < 0$, N is the number of spins in the atomic plane parallel to the atomically smooth interface, the distance is measured in units of the distance between the two nearest such planes, and the dimensionless constants $\alpha_{f(\mathrm{af})}$ and β are equal, respectively

$$\alpha_{f(\mathrm{af})} = 2K_{||}\Big/ z_2\left|J_{f(\mathrm{af})}\right| \ll 1, \tag{5.5}$$

$$\beta = \mu_{\mathrm{at}} B_0 \Big/ z_2 J_f, \tag{5.6}$$

where μ_{at} is the magnetic moment of the ferromagnet atom and B_0 is the induction of an external magnetic field.

In the case of weak anisotropy $\left(K_{||} \ll \min\left(J_f, |J_{\mathrm{af}}|\right)\right)$, we can assume that at the interface

$$\theta_f(0) - \theta_{\mathrm{af}}(0) = \frac{\pi}{2}. \tag{5.7}$$

Without limiting the generality of the consideration, we choose the numbering of the two sublattices of the collinear antiferromagnet so that, far from the interface of the layers, the magnetization vector of the ferromagnet is parallel and not antiparallel to the antiferromagnetism vector. Then, their turn from the axis of easy magnetization occurs in different directions. Let, for certainty, $\theta_f > 0$, and $\theta_{\mathrm{af}} < 0$.

By varying the energy (5.4), we obtain the following Euler equations

$$\theta''_{\mathrm{af}} = \alpha_{\mathrm{af}} \sin 2\theta_{\mathrm{af}} \tag{5.8}$$

and

$$\theta''_f = \alpha_f \sin 2\theta_f + \beta \sin \theta_f. \tag{5.9}$$

The solution of the first of them has the form (see L. D. Landau, E. M. Lifshits. *Electrodynamics of continuous media* (Moscow: Nauka, 1982), §43)

$$\cos\theta_{\mathrm{af}}(x) = \mathrm{th}\left[-(2\alpha_{\mathrm{af}})^{\frac{1}{2}}(x + x_1)\right], \tag{5.10}$$

and the second one

$$\sin\theta_f(x) = \frac{2[\beta(2\alpha_f + \beta)]^{1/2} \mathrm{sh}\left[(2\alpha_f + \beta)^{\frac{1}{2}}(x + x_2)\right]}{2\alpha_f + \beta \mathrm{ch}^2\left[(2a_f + \beta)^{\frac{1}{2}}(x + x_2)\right]}. \tag{5.11}$$

The constants x_1 and x_2 are expressed by $\theta_f(0)$ and $\theta_{\mathrm{af}}(0)$. Using the ratio (5.7) and minimizing the total energy of the spiral by the remaining parameter, we find [2]

$$\cos\theta_f(0) = \frac{\varepsilon_f\left\{\left[\varepsilon_f^2\beta^2 + 4\alpha_f(\alpha_f+\beta)\left(\varepsilon_f^2+\varepsilon_{\mathrm{af}}^2\right)\right]^{1/2} - \varepsilon_f\beta\right\}}{2\alpha_f\left(\varepsilon_f^2+\varepsilon_{\mathrm{af}}^2\right)}, \tag{5.12}$$

where $\varepsilon_f \propto \alpha_f^{1/2}$ and $\varepsilon_{\mathrm{af}} \propto \gamma_{\mathrm{af}}\alpha_{\mathrm{af}}^{1/2}$ are the surface energies of the domain walls in the corresponding layers in the absence of an external field.

By $B_0 = 0$

$$tg\theta_f(0) = \frac{\varepsilon_{\mathrm{af}}}{\varepsilon_f}, \tag{5.13}$$

The spiral is mainly located in the layer in which the surface energy of the domain wall is lower.

As the external field increases, the spiral is pushed into the antiferromagnetic layer. By $\beta \gg \alpha_f$

$$\theta_f(0) = \frac{\varepsilon_{\mathrm{af}}}{\varepsilon_f}\sqrt{\frac{2\alpha_f}{\beta}} \propto \beta^{-1/2}. \tag{5.14}$$

In deriving this formula, we neglected the bevel of the sublattices of the antiferromagnet, assuming that the external field is much smaller than the spin-flop transition field in the antiferromagnet.

When the ferromagnet is remagnetized, the direction of the exchange helix twisting changes to the opposite. In this case, the energy of the interface does not change; therefore, there is no exchange shift in this state.

5.2 The Case of "Charged" Edges of Atomic Steps at the Interface "Ferromagnet–Antiferromagnet"

Consider a simple cubic lattice of a collinear two-lattice antiferromagnet whose planes (001) are parallel to the interface and perpendicular to the axis z of the Cartesian coordinate system. The axes x and y of this system are parallel to the directions [110] and [10]. The light axes in the plane (001) are similar to the directions [100] and [010]. The crystal lattice of a ferromagnet is a continuation of the crystal lattice of an antiferromagnet; the difference in the lattice constants is negligible.

Consider an atomic step parallel to the axis y. As can be seen from Fig. 5.3, it leads to the appearance of an uncompensated spin on the surface of the antiferromagnet, parallel to the direction of magnetization of one of the sublattices.

The presence of uncompensated spins at the edge of the stage leads to a distortion of the homogeneous distribution of the magnetic parameters of the order.

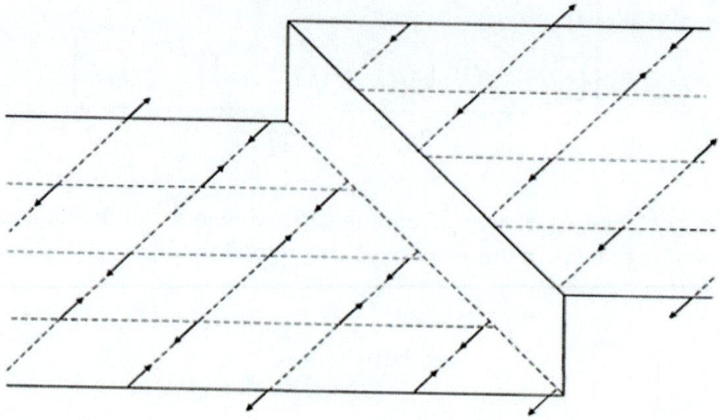

Fig. 5.3 Uncompensated spins of the antiferromagnet at the boundary of the atomic step

Let us consider successively two limiting cases of weak and strong roughness.

5.2.1 Case of Weak Roughness

In this case, the characteristic distance between the parallel edges of the atomic steps R is quite large and exceeds the widths $\Delta_{f(\mathrm{af})}$ of the traditional (Bloch or Neel) domain walls in magnetic materials.

In the case of local distortions of the magnetic order parameters, the distortions occurring near the edges of neighboring atomic steps do not overlap, and it is possible to solve the problem of inhomogeneity of the magnetic order parameters near the edge of a separate step [3].

The non-locality of distortions can be related, according to the Malozemov model [A. P. Malozemov. *Phys. Rev. B* **35,** 3679 (1987)], with the appearance of bubble-like domains (Fig. 5.4).

Indeed, the exchange field created by uncompensated spins is directed along with one of the sublattices of the antiferromagnet; that is, at a slight angle of the sublattice slope, it is almost perpendicular to the magnetization vector of the ferromagnet at the spin-flop orientation of the order parameters. Therefore, there is a possibility that uncompensated spins can cause the appearance of a 180° domain wall near the edge

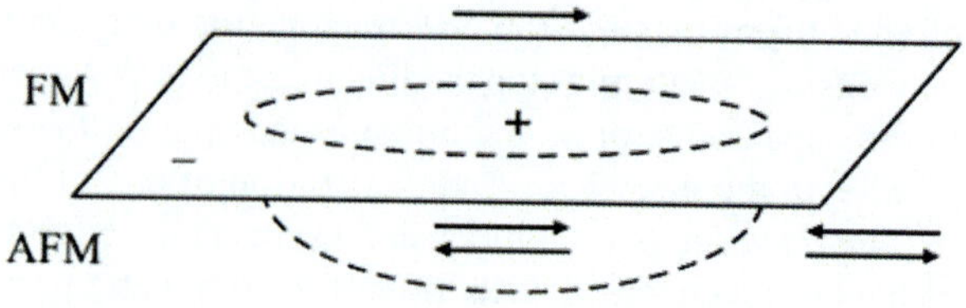

Fig. 5.4 Schematic representation of a bubble-shaped domain

of the step, the middle of which coincides with the edge of the step. In this case, the orientation of the uncompensated spins of the antiferromagnet near the edge of the step will be parallel (at $J_{f,\text{af}} > 0$) to the magnetization of the ferromagnet.

In the case of a thin ferromagnetic layer, it is energetically more advantageous than splitting into bubble-like domains to split the ferromagnetic layer into domains by walls perpendicular to the plane of the layer, and the midpoints of which coincide with the edges of the steps.

However, as the results of our simulation show, there is no partitioning into domains. The minimum energy corresponds to a state with local distortions of the magnetic order parameters near the edges of the atomic steps. The distortions are mainly concentrated in the layer with lower exchange stiffness.

The energy disadvantage of the appearance of bubble-like domains can be seen from the following simple estimates. In the formation of such domains, the loss in energy due to the formation of a domain bounded by two steps, calculated per unit length of the step, is of the order of $\varepsilon_{f(\text{af})}R$, where

$$\varepsilon_{f(\text{af})} \sim \sqrt{J_{f(\text{af})}K_{\|}}\Big/ b^2 \sim \frac{J_{f(\text{af})}}{\Delta_{f(\text{af})}b^2}. \tag{5.15}$$

The wall occurs in a layer with a lower value $\varepsilon_{f(\text{af})}$.

The gain in energy due to the parallelism of the uncompensated spins of the antiferromagnet and the magnetization of the ferromagnet will amount the value $J_{f,\text{af}}\big/ b$ to. At $R > \Delta_{f(\text{af})}$ the same time, the loss in energy exceeds the gain, so the formation of bubble-like domains does not occur.

It would seem that by increasing $J_{f,\text{af}}$, we can make the state with domains energetically profitable. However, this is an illusion, since at $J_{f,\text{af}} > \min\left(J_f, |J_{\text{af}}|\right)$ at the interface of the layers, the spins belonging to different layers remain collinear. The increased energy will have an exchange bond between the spins of the first and second (counting from the interface) atomic planes belonging to a magnet with a lower value of exchange stiffness. Thus, the lowest of the exchange energies will be included in the estimate.

5.2.2 *A Case of Strong Roughness. Thick Layer*

We now investigate the case $R \ll \Delta_f, \Delta_{\text{af}}$ when the effect of the "charges" of the atomic steps on the magnetic order parameters is collective. As will be shown below, in this case, under certain conditions, it is possible to destroy the long-range order near the interface of the layers caused by the action of random fields.

This phenomenon has already been discussed in the theory of phase transitions. As shown in the paper by Imry and Ma [Y. Imry, S-k. Ma. Phys. Rev. Lett. **35**, 1399 (1975)], in the absence of anisotropy, arbitrarily weak defects of the "random local field" type lead to the disappearance of the long-range order. In the presence of weak anisotropy, this disappearance has a threshold character; that is, it occurs when the concentration of defects (or their "strength") exceeds a particular critical value.

The specificity of our problem is that the edges of the atomic steps, i.e., linear defects of the "random local field" type, are concentrated at the interface of the magnetic layers. Therefore, the long-range magnetic order disappearance will occur in the area of finite thickness adjacent to the interface and located in a layer with a lower exchange stiffness.

Find the condition for the appearance of such a "broken" layer. Let the characteristic size of the inhomogeneity of the order parameter in the layer be the value $L \gg R$. At the same time L, is less than the layer thickness. Since the sign of the random field occurring at the edge of the atomic step is random, the fluctuation of the random field along the length L in the direction perpendicular to the edges of the atomic steps is of the order of magnitude $\frac{h_0}{b}\sqrt{\frac{L}{R}}$ per unit length of the stage. The corresponding contribution to the surface energy density due to the local parallelism of the order parameter and the random field fluctuation will be [2]

$$w_1 \sim -\frac{h_0}{b\sqrt{LR}}. \tag{5.16}$$

The loss in the surface energy density due to the inhomogeneity of the order parameter in order of magnitude is equal to

$$w_2 \sim \frac{J}{Lb}, \tag{5.17}$$

since the characteristic size of the area of inhomogeneity in the direction perpendicular to the interface is also of order L.

The increase in the surface anisotropy energy density is

$$w_3 \sim \frac{KL}{b^3}. \tag{5.18}$$

If $w_3 > |w_1|$, then the occurrence of inhomogeneity on large (compared to R) scales is not energetically advantageous. The condition $|w_1| > w_3$ leads to the inequality

$$L^{3/2} < \frac{h_0 \Delta^2}{J R^{1/2}}, \tag{5.19}$$

where Δ is the width of the traditional domain wall in the layer with lower exchange stiffness.

Neglecting w_3, we find the optimal value L^* corresponding to the minimum of the total energy $w_1 + w_3$

$$L^* \sim \left(\frac{J}{h_0}\right)^2 R. \tag{5.20}$$

Since the distance L^* should be much greater than the parameter R, it can be seen that the "broken" layer occurs only when the condition is met

$$\left|J_{f,\text{af}}\right| \ll \min\left(|J_{\text{af}}|, J_f\right). \tag{5.21}$$

Substituting L^* in the inequality (5.19), we get $L^* < \Delta$ that the scale of inhomogeneities is less than the width of the domain wall, and the domains do not have time to form on this scale. Thus, bubble-like domains do not occur in the case of either weak or strong roughness.

The condition $|w_1| > w_3$ is equivalent to the following restriction on the strength and concentration of defects of the "random local field" type:

$$\frac{h_0^2}{R} > \frac{J^2}{\Delta}. \tag{5.22}$$

5.2.3 A Case of Strong Roughness. Thin Layer

If the thickness of a thin (ferromagnetic) layer d is less than the critical value d^*, then the area of the destroyed long-range order will occupy the entire volume of this layer. In the case $d \ll d^*$, the problem becomes one-dimensional since the change in the order parameter over the thickness of the thin layer can be ignored.

If the thin layer is a layer with less exchange stiffness, then $d^* \sim L^*$.

Consider the case when the area of the destroyed long-range order is concentrated in a thin layer. Then

$$w_2 \sim \frac{Jd}{bL^2}, \tag{5.23}$$

$$w_3 \sim \frac{Kd}{b^3}. \tag{5.24}$$

The condition $|w_1| > w_3$ gives [3]

$$L < \left(\frac{h_0}{J}\right)^2 \frac{\Delta^4}{d^2 R} \tag{5.25}$$

In this case, the optimal value of the inhomogeneity scale is

$$\tilde{L} \sim \left(\frac{J}{h_0}\right)^{2/3} (d^2 R)^{1/3}. \tag{5.26}$$

The inequality (5.25) is equivalent to the condition $\tilde{L} < \Delta$, and the constraint $\tilde{L} \gg R$ gives

$$Jd \gg Rh_0. \tag{5.27}$$

The condition for the occurrence of a "broken" layer, similar to the inequality (5.22), has the form

$$\frac{h_0^2}{R} > \frac{J^2 d^2}{\Delta^3}. \tag{5.28}$$

In the case when the exchange stiffness of a thin layer is higher than that of a thick layer, equating the total surface energy densities for the considered cases of thick and thin layers, we find the following critical value of the thickness d^*

$$d^* \sim \left(\frac{|J_{\mathrm{af}}|}{J_f}\right) L^*. \tag{5.29}$$

At $d > d^*$, the "broken layer" is located in a softer substrate, and at $d \sim d^*$, it passes into a more stiff but thinner film as the film thickness decreases.

It is easy to show that in the case of a square grid of the edges of atomic steps with a characteristic size R, conducting a similar consideration, the conditions for the occurrence of a "broken" layer are not realized.

Thus, in weak random fields, as a result of their collective influence, a layer with a disturbed long-range magnetic order is formed near the film-substrate interface under certain conditions. If the film thickness is less than the critical value, this layer occupies the entire volume of the film.

References

1. Morosov AI (2003) Magnetic structure of the compensated ferromagnet-antiferromagnet interface. FTT 45(10):1847–1849
2. Morosov AI, Markets DO (2007) Magnetic structure of the ferromagnet–antiferromagnet interface with parallel anisotropy axes. FTT 49(10):1849–1852
3. Morosov AI, Morosov IA, Sigov AS (2010) Distortions of the magnetic parameters of the order caused by the "charged" edges of the atomic steps at the ferromagnet-antiferromagnet interface. FTT 52(2):302–306

Chapter 6
Behavior in a Magnetic Field

6.1 Exchange Bias. Uncompensated Slice

Let us now consider the possible mechanisms of the exchange bias of the hysteresis loop of a ferromagnetic film due to the influence of an antiferromagnetic substrate, which was discussed in the first chapter of the book (Sect. 1.2). The first model of the exchange bias for an uncompensated slice of an antiferromagnet, proposed in the paper [W. H. Meiklejohn, C. P. Bean. *Phys. Rev.* **102**, 1413 (1956)], assumed a break in the exchange bonds at the ferromagnet-antiferromagnet interface and gave an exchange bias value about hundred times higher than the observed values.

A more realistic model by Mowry et al. [D. *J. Appl. Phys.* **62**, 3047 (1987)] proceeds from the assumption that a magntization reversal of a ferromagnetic film leads to creation of a 180° domain wall (exchange spiral) in an antiferromagnetic substrate near an atomically smooth interface. According to this model, the magnetic induction B_{ex}^{af} that characterizes the exchange bias can be obtained from the equality of the difference between the Zeeman energies before and after the formation of the exchange helix and the energy of the domain wall [1]:

$$B_{\mathrm{ex}}^{\mathrm{af}} = \frac{\varepsilon_{\mathrm{af}}}{2M_{\mathrm{f}}ab} \sim \frac{\left(J_{\mathrm{af}}K_{\parallel}\right)^{1/2}}{b^3 M_{\mathrm{f}}a}, \tag{6.1}$$

where M_f is the magnetization of the ferromagnet.

It should be noted that the state with the exchange spiral does not meet the global power minimum: it is advantageous for the domain wall to pass through the anti-ferromagnet. In this case, the antiferromagnet will pass into a homogeneous state. That is prevented by pinning the domain wall by defects in the crystal structure of the antiferromagnet, which is the one that causes the exchange bias.

However, it is necessary to pay attention to the fact that the exchange spiral occurs in a layer with a lower surface power of the Bloch domain wall and, under certain conditions, can also occur in a ferromagnetic layer, and then, with an increase in the external magnetic field, pass into an antiferromagnet. In this case, it is necessary to

A. S. Sigov, *Multilayer Magnetic Nanostructures*, Springer Aerospace Technology,
https://doi.org/10.1007/978-981-19-6246-2_6

consider the dependence of the thickness and surface power of the domain wall on the magnetic field [S. Mangin, G. Marchal, B. Barbara. *Phys. Rev. Lett.* **82**, 4336 (1999)]

$$\Delta_{\mathrm{f}}(B_0) \sim \left(\frac{J_{\mathrm{f}}}{K_{\parallel} + 2M_{\mathrm{f}} B_0 b^3} \right)^{1/2}, \tag{6.2}$$

$$\varepsilon_{\mathrm{f}}(B_0) \sim \left[J_{\mathrm{f}} \left(K_{\parallel} + 2M_{\mathrm{f}} B_0 b^3 \right) \right]^{1/2} / b^2, \tag{6.3}$$

where B_0 is the magnetic field induction.

In the case $\varepsilon_{\mathrm{f}}(0) < \varepsilon_{\mathrm{af}}$, a domain wall in a ferromagnetic film can only occur under the condition $\Delta(B_0) < a$. Equating the power of the domain wall to the gain in the Zeeman power, we obtain the exchange bias

$$B_{\mathrm{ex}}^{\mathrm{f}} \approx \begin{cases} \left(J_{\mathrm{f}} K_{\parallel} \right)^{1/2} / M_{\mathrm{f}} a b^3, & a \gg \Delta_{\mathrm{f}}(0), \\ \frac{J_{\mathrm{f}}}{M_{\mathrm{f}} a^2 b^3}, & a \ll \Delta_{\mathrm{f}}(0), \end{cases} \approx \frac{J_{\mathrm{f}}}{M_{\mathrm{f}} a b^3 \min(a, \Delta_{\mathrm{f}}(0))}. \tag{6.4}$$

With the growth of the magnetic field, the domain wall becomes thinner and in the field $B_0 \sim B^*$ passes from the ferromagnet to the antiferromagnetic substrate, where it remains at $B_0 > B^*$. The value B^* is found from the equality of the surface energies of the Bloch domain walls in the layers:

$$B^* \sim \frac{|J_{\mathrm{af}}| K_{\parallel}}{J_{\mathrm{f}} M_{\mathrm{f}} b^3}. \tag{6.5}$$

If $B_{\mathrm{ex}}^{\mathrm{f}} > B_{\mathrm{ex}}^{\mathrm{af}}$, then the domain wall arises immediately in the substrate. The dependence of the exchange bias on the film thickness is shown in Fig. 6.1.

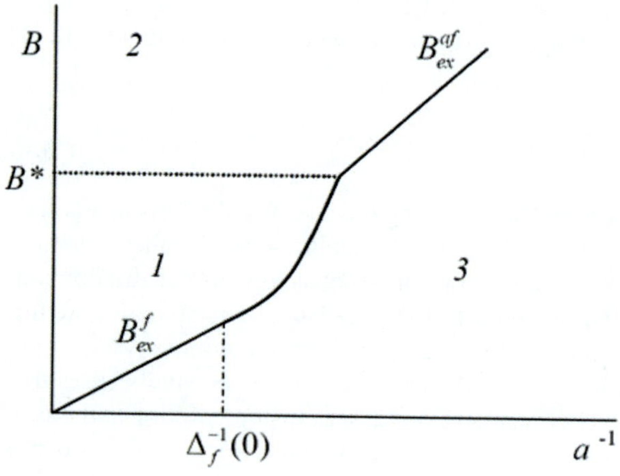

Fig. 6.1 The phase diagram "magnetic induction—film thickness" in the case of a smooth interface: 1—phase with a domain wall in a ferromagnet, 2—phase with a domain wall in an antiferromagnet, 3—phase without a domain wall. The solid line shows the dependence of the exchange shift on the inverse thickness of the film

Mathematical modeling has shown that in the field $B^{\text{f}}_{\text{ex}} < B < B^*$, there is a bound state of the domain wall at the interface of the layers. In fields with induction $B_0 > B^*$, it is absent, and the exchange bias is due to the pinning of the domain wall.

Let us now consider the phenomenon of exchange bias in the case of a rough surface in different areas of the phase diagram "film thickness-roughness". There is no exchange bias in the polydomain phase, since, at one direction of the field in saturation, exchange spirals are formed in areas of one type and at the opposite direction of the field, they are formed in areas of another type. In this case, the energies of states with opposite directions of magnetization are the same.

In an external magnetic field whose direction is parallel (domains of the first type) or antiparallel (domains of the second type) to the magnetizations in the domains, the magnetization in the domains of the first type does not change. In the domains of the second type, the magnetization reverses by 180^0. If the size of the domain R exceeds the thickness of the traditional domain wall in an antiferromagnet Δ_{af}, then this reversal is accompanied by the appearance of such a wall in the substrate near the interface and the characteristic magnetization field is equal in the order of magnitude B^0_{af}.

If $R \ll \Delta_{af}$, then the reversal of the magnetization in the domain is accompanied by the appearance of a static spin vortex in the antiferromagnetic substrate, the characteristic magnetization field is on the order of magnitude $B^0_{\text{af}}\Delta_{\text{af}}/R$. At the same time, a 90° domain wall appears in the substrate. The reasons for its occurrence are similar to the reasons why, in the absence of a field, the magnetization of the film in the monodomain phase is orthogonal to the order parameter in the depth of the substrate: its occurrence reduces the power of the vortex system [2].

Indeed, in the absence of a domain wall in those domains where the magnetization is parallel to the external field, no vortices are formed, and in domains with an initially antiparallel orientation of the magnetization and the field $B^0_{\text{af}}\Delta_{\text{af}}/R$, a 180° vortex occurs in the field. In the presence of wall, 90° vortices occur in both domains, and the rotation of the antiferromagnetic order parameter in different types of domains occurs in opposite directions. Since the power of the vortex is proportional to the square of the angle of rotation, in the presence of a wall, the power of the vortices is lower, and the power difference between the states without a domain wall and with a wall exceeds the power of its formation.

If the magnetic field is applied in the plane of the film perpendicular to the magnetizations of the domains, the characteristic magnetization field has the same order of magnitude as in the case of a field parallel to the magnetizations. However, the 90° domain wall does not occur since 90° static vortices with the opposite direction of rotation of the antiferromagnetic order parameter happen in both types of domains.

Similarly, there is no exchange bias in the phase with 180° exchange spirals parallel to the interface boundaries. This phase is realized in the case of one easy axis in the plane of the layer (phase 4 in Fig. 4.6a and b). Indeed, when the ferromagnetic layer is remagnetized, these spirals disappear in areas of one type where they were present before the remagnetization and arise in areas of another type.

In the case of one easy axis in the layer plane, the exchange bias should not occur in the phase with vortices and 90° exchange spiral (phase 3 in Fig. 4.6a and b). Remagnetization of the ferromagnet will only lead to a change in the direction of rotation in the exchange spiral.

Thus, in the case of one easy axis in the plane of the layer, the exchange bias occurs only in the phase with vortices and the magnetization of the ferromagnet perpendicular to the easy axis (phase 2 in Fig. 4.6a and phases 4 and 5 in Fig. 4.6b).

In the case of two mutually perpendicular easy axes in the plane of the layer, the exchange shift is present in the entire area of existence of the monodomain phase of the ferromagnetic film. Therefore, it can be concluded that the presence of two mutually perpendicular easy axes in the plane of the layer contributes to the appearance of an exchange bias.

6.2 Exchange Bias. Compensated Slice

The first mechanism of the occurrence of the exchange bias in the case when the spins of the atomic plane of the antiferromagnet parallel to the interface are compensated is considered in the paper by Malozemov [A. P. Malozemov. *Phys. Rev. B* **35,** 3679 (1987)]. It is assumed that the roughness of the interface leads to the appearance of uncompensated spins of the antiferromagnet on it. The exchange interaction of uncompensated spins of the antiferromagnet with the spins of the ferromagnet leads to the formation of local random fields conjugated to the antiferromagnetic order parameter, and a random field $\pm h_0 \left(h_0 \approx J_{\mathrm{f,\,af}}\right)$ occurs in each unit cell at the interface. The characteristic fluctuation of the surface exchange power averaged over the area L^2 is $h \approx \pm h_0/\sqrt{N}$, where N is the number of elementary cells in the specified area.

In this random exchange field, under the condition of the monodomain state of the ferromagnet, the surface layer of the antiferromagnet will split into bubble-like domain-like areas with a characteristic size of the order of the width of the traditional domain wall (Fig. 5.4), to reduce the total power of the system. The resulting exchange bias in the order of magnitude is given by the formula (6.1).

However, the assumption of the occurrence of a random field in each unit cell is not justified. Indeed, let the atomic steps at the interface go in two mutually perpendicular directions x and y. Then, the entire interface is divided by them into polygons with right angles. By continuing one of the sides (e.g., parallel to the x-axis) of each inner corner equal to $3\pi/2$, you can split the polygon into rectangles. If the lengths of both sides of such a rectangle l_x and l_y are an odd number of lattice constants, that is, the number of elementary cells in it is odd, then in such a rectangle, there is no compensation for the fields created by the spins of the near-surface atomic layer of the film. Thus, the random field does not occur in every cell, but only in the characteristic area $S_0 = 4\left(l_x^{-1} l_y^{-1}\right)^{-1} \sim R^2$, where the brackets denote the averaging over the interface, and the multiplier 4 occurs under the assumption that for $R \gg 1$, the length of the step is equally likely equal to an even or odd number of lattice constants.

The characteristic fluctuation of the field in an area with a linear size L will be calculated based on an elementary cell of the order of [3]

$$\langle \tilde{h} \rangle \sim h_0 \left(\frac{L}{R} \right) / L^2 \sim \frac{h_0}{RL}, \tag{6.6}$$

which is R times smaller than the one introduced in Malozemov's work.

However, even in the case of a R-fold overestimation of the gain in the power of the film-substrate boundary due to the appearance of domains in the substrate, the total power of the system at $J_{af} \approx J_{f,\,af}$ is higher than in the case of a homogeneous distribution of the ferromagnetic and antiferromagnetic order parameters. That is, the phase with domains in the substrate is metastable. That is all the more true in the case of an actual estimate of the gain in surface power.

Thus, in the case of a compensated surface of an antiferromagnet, the proposed mechanism for the occurrence of an exchange bias is not realized.

As shown in [U. Nowak et al. *Phys. Rev. B* **66**, 014, 430 (2002)], the introduction of non-magnetic atoms of a sufficiently large concentration leads to the formation of magnetic clusters with uncompensated spins that are weakly bound to the antiferromagnet matrix, and the freezing of which, when cooled in a magnetic field, creates an exchange bias. However, the consideration of highly disordered systems is beyond the scope of this book.

In the model proposed by Koon [N. C. Koon. *Phys. Rev. Lett.* **78**, 4865 (1997)], the degree of roughness of the interface does not play a significant role, and the exchange is due to the spin-flop orientation of the order parameters at the interface. As in the case of an uncompensated surface, when a ferromagnet is remagnetized, a 180^0 domain wall appears in the antiferromagnet. As shown above, in the case of a single easy axis in the plane of the layer, the presence of an exchange spiral in the absence of a field eliminates the exchange bias. It will take place either in the area of applicability of the exchange approximation $a \ll \min(\Delta_f, \Delta_{af})$, or in the presence of two mutually perpendicular easy axes in the plane of the layer. In these cases, the exchange spiral is absent in the initial state and occurs in remagnetization. Therefore, as in an uncompensated surface, this configuration of easy axes contributes to the appearance of an exchange in the ferromagnet-antiferromagnet system.

In a later work [T. C. Schulthess and W. H. Butler. *Phys. Rev. Lett.* **81**, 4516 (1998)], the authors came to a result that is the opposite of the conclusion of Koon. They considered the same model and performed mathematical simulations using the Landau-Lifshitz equations to describe the motion of magnetic moments. The authors found the presence of a spin-flop orientation but found no exchange bias. Their calculations show that an antiferromagnetic layer only leads to an increase in coercive force. They explained this difference by adding a degree of freedom, that is, by allowing the spins of the antiferromagnet to escape from the plane parallel to the interface. That leads to a decrease in the power barrier that the antiferromagnet spins need to overcome to move from the metastable state with the domain wall to the ground state.

This explanation seems to us to be incorrect. The fact is that a potential barrier between the metastable and the ground state exists even in the presence of two degrees of freedom of spins. To overcome it, when modeling the process of remagnetization in a quasi-statically changing magnetic field, a term describing thermal excitation, for example, Langevin noise, must be introduced into the equations of motion. However, there is no mention of such a term in work. Without it, the system cannot overcome the barrier. The absence of the exchange displacement can be explained by the passage of the domain wall through the entire layer of the antiferromagnet, as discussed above. Indeed, with the values of the exchange parameters used in the works of Koon and his critics, the bound state of the domain wall at the interface of the layers is absent, and any defects that can fix the domain wall were not introduced into the model.

6.3 A Substrate of Finite Thickness. "Switching" the Nanodomain State

Consider the behavior of the polydomain phase of a two-layer ferromagnet-antiferromagnet system in the case when the exchange stiffness of the ferromagnet is lower than that of the antiferromagnet. When the thickness of the layers is approximately equal, when the characteristic distance between the adjacent edges of the steps at the interface R exceeds the thickness of the layers, the ferromagnetic layer is divided into 180° domains due to frustrations. Domain walls of a new type, generated by frustrations, originate at the edges of the atomic steps and permeate the ferromagnetic layer, expanding as they move away from the interface (Fig. 4.1).

Such a state is stable in a zero or weak external magnetic field. The external field applied collinearly to the magnetizations of the domains, exceeding the critical value, makes the monodomain state of the ferromagnetic layer more advantageous. In this state, the domain walls, which have a more complex and field-dependent structure, divide the antiferromagnetic layer into 180° domains.

We will limit ourselves to the area of applicability of the exchange approximation when all the characteristic geometric dimensions: layer thicknesses and widths of the domain walls resulting from frustrations are much smaller than the widths of traditional (Bloch, Neel, or hybrid) domain walls, determined by the competition of the exchange power and the anisotropy power.

The power of the domain walls in the antiferromagnetic layer is higher than in the ferromagnetic layer. The value of the critical magnetic field is found from the conditions of equality of the gain in the Zeeman power and the loss in the power of the domain walls. Therefore, to find the dependence of the "switching" field on the system parameters, it is necessary to first study the structure of the domain walls in the antiferromagnet.

To the energies described in Sect. 4.1, the power of interaction with an external magnetic field (the Zeeman power) was added [4]

$$W_Z = \mu \sum_i (s_i, B), \tag{6.7}$$

where μ is the magnetic moment of the unit cell and B is the induction of an external magnetic field. Here, it is taken into account that the spin vector is antiparallel to the magnetic moment vector of the atom.

Since the magnetic phase diagram is determined by the characteristic R distance between the edges of the atomic steps, replacing the parallel steps with their square grid with the same cell size R does not lead to a qualitative change in the resulting phase diagram.

6.3.1 *The Area of Strong Fields*

As shown by the simulation, in the area of strong fields, the domain wall generated by the atomic step in the antiferromagnet has the structure shown in Fig. 6.2.

The spins of the atoms lie in the plane of the layers; the order parameter forms an angle θ with an easy axis. The area of distortion of the order parameter in a ferromagnet has an oval shape with characteristic dimensions a and δ (Fig. 6.2a). The width of the domain wall in the antiferromagnetic layer increases as it moves away from the interface of the layers from the value δ to the value of the order of the thickness of the antiferromagnetic layer d_{af}. The dependence of the parameters a and δ the magnitude of the magnetic field can be found from simple power estimates.

The power of the expanding domain wall in the area of the ferromagnet adjacent to the edge of the step, in order of magnitude (per unit length of the edge of the atomic step), has the form

$$w_1 \sim \frac{J_f}{b} \ln a, \tag{6.8}$$

where b is the interatomic distance.

Similarly, the power of the expanding domain wall in an antiferromagnet is equal to

$$w_2 \sim \frac{|J_{\mathrm{af}}|}{b} \ln \frac{d_{\mathrm{af}} + \delta}{\delta}. \tag{6.9}$$

The power of the inhomogeneity of the order parameter in the area of semiovals in a ferromagnet in order of magnitude is

$$w_3 \sim \frac{J_{\mathrm{f}} a \delta}{b} \left(\frac{1}{a^2} + \frac{1}{\delta^2} \right). \tag{6.10}$$

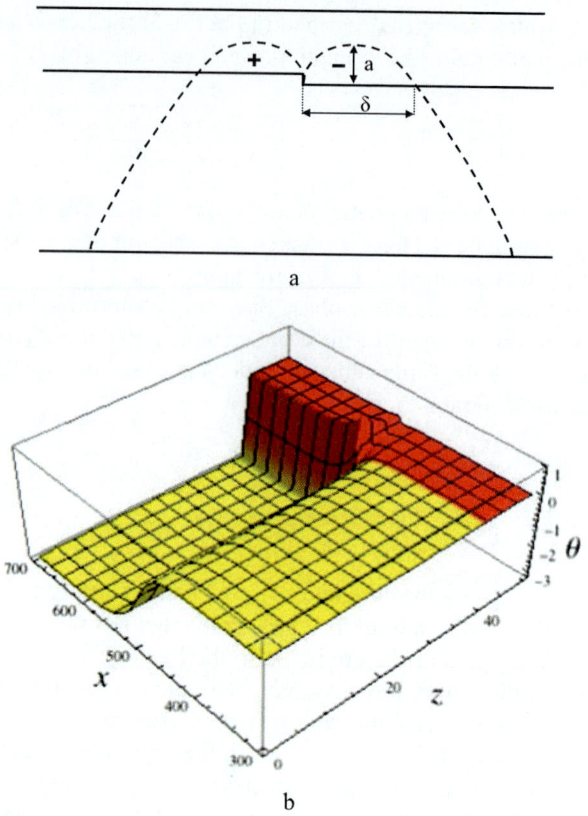

Fig. 6.2 The domain wall generated by the atomic step in the area of strong ($h = 0.03$) magnetic fields: schematic representation (**a**) and the angle θ calculated for the case $J_{\rm af} = -1$, $J_f = J_{\rm f,af} = 0, 1$ (**b**). The step at the interface between the ferromagnet ($z > 40$) and the antiferromagnet ($z < 40$) corresponds to the coordinate $x = 500$

The loss in power, because in the area of semi-intervals, the order parameter is non-collinear to the external magnetic field, is equal to

$$w_4 \sim \frac{ha\delta}{b}, \tag{6.11}$$

where $h = \mu B$ is the Zeeman power per unit cell.

Minimizing the total power $w_1 + w_2 + w_3 + w_4$ by the parameters a and δ, we find in the case of $d_{af} \gg \delta$

$$a \sim \sqrt{\frac{J_{\rm f}}{h}}, \tag{6.12}$$

$$\delta \sim \frac{|J_{\rm af}|}{\sqrt{J_{\rm f}h}}. \tag{6.13}$$

This solution holds if $a \ll d_{\rm f}$ and $\delta \ll d_{\rm af}$. The first of these inequalities is violated in the field

$$h_1 = J_f d_f^{-2}, \tag{6.14}$$

and the second—in the field

$$h_2 = \frac{J_{af}^2}{J_f d_{af}^2}. \tag{6.15}$$

6.3.2 The Area of Medium Fields

If $h_2 > h_1$, which is equivalent to the inequality

$$d_f|J_{af}| > d_{af} J_f, \tag{6.16}$$

then, in the area $h_2 > h > h_1$, the minimum power is reached at a, given by the formula (6.12), and

$$\delta \sim \sqrt{\frac{d_{af} a |J_{af}|}{J_f}} > d_{af}. \tag{6.17}$$

We will limit ourselves to considering the case $d_{af}|J_{af}| > d_f J_f$ when $\delta > a$.

If $h_1 > h_2$, then at $h < h_1$ the domain wall has the form shown in Fig. 6.3.

The area of inhomogeneity of the ferromagnetic order parameter occupies the entire thickness of the ferromagnetic layer (Fig. 6.3a). Then, the inhomogeneity power of the order parameter in a ferromagnet has the form [5]

$$w_1 + w_3 \sim \frac{J_f d_f}{b\delta} + J_f \ln d_f, \tag{6.18}$$

and the increase in the Zeeman power

$$w_4 \sim \frac{h d_f \delta}{b}. \tag{6.19}$$

Minimizing the power by the parameter δ under the assumption $\delta \ll d_{af}$, we obtain

$$\delta \sim \frac{|J_{af}|}{h d_f}, \tag{6.20}$$

and the critical value h_2^* at which $\delta = d_{af}$, takes the form

$$h_2^* = \frac{|J_{af}|}{d_f d_{af}}. \tag{6.21}$$

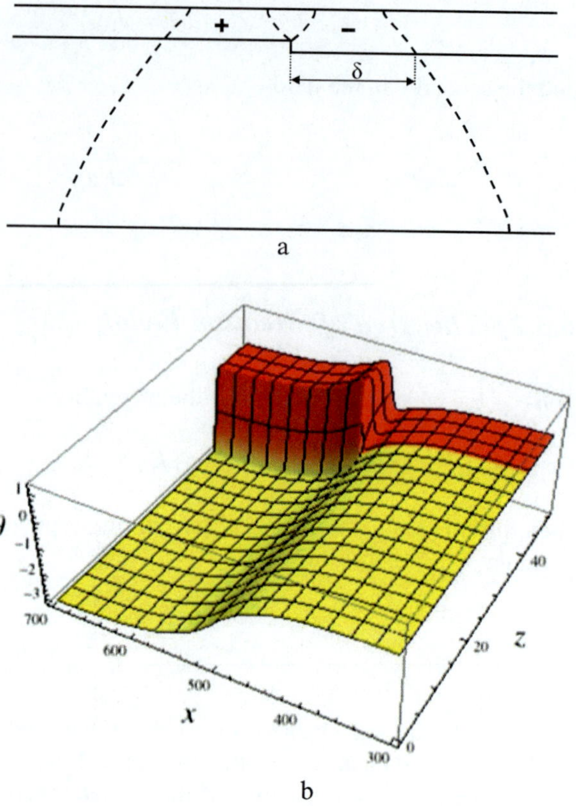

Fig. 6.3 The domain wall generated by the atomic step in the area of average ($h = 0.002$) magnetic fields: schematic representation (**a**) and the angle θ calculated for the case $J_{\mathrm{af}} = -1$, $J_{\mathrm{f}} = J_{\mathrm{f,af}} = 0.1$ (**b**). The step at the interface between the ferromagnet ($z > 40$) and the antiferromagnet ($z < 40$) corresponds to the coordinate $x = 500$

6.3.3 *Weak Field Area*

In the field range $h < \min(h_1, h_2)$, the parameter $\delta > d_{\mathrm{af}}$ and the inhomogeneity area occupies the ferromagnetic layer's entire thickness (Fig. 6.4).

The power of the domain wall in an antiferromagnet has the form

$$w_2 \sim |J_{\mathrm{af}}| \frac{d_{\mathrm{af}}}{b\delta}. \tag{6.22}$$

The power of the order parameter inhogeneity in a ferromagnet and the loss in the Zeeman power are given by formulas (6.18) and (6.19), respectively.

Minimizing the total power leads to the value

$$\delta(h) \sim \sqrt{\frac{|J_{\mathrm{af}}| d_{\mathrm{af}}}{h d_{\mathrm{f}}}} \gg d_{\mathrm{f}}, d_{\mathrm{af}}. \tag{6.23}$$

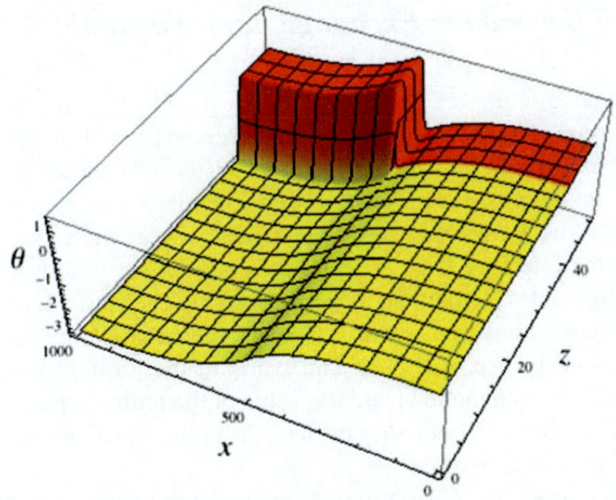

Fig. 6.4 The domain wall generated by the atomic step in the range of weak ($h = 0.0001$) magnetic fields: the angle θ calculated for the case $J_{\text{af}} = -1, J_{\text{f}} = J_{\text{f, af}} = 0.1$. The step at the interface between the ferromagnet ($z > 40$) and the antiferromagnet ($z < 40$) corresponds to the coordinate $x = 500$

6.3.4 The "Switching" Field

It is easy to see that according to the formula (6.23), with a decrease in the magnitude of the magnetic field, the parameter δ increases. The simulation shows that the "switching" occurs gradually and is not, strictly speaking, a phase transition. The "switching" h_0 field corresponds to the value [6]

$$\delta(h_0) = R, \tag{6.24}$$

The "switching" field is the field in which the domain walls occupy the entire space between the atomic steps and begin to overlap.

From condition (6.24), we get an estimate for the "switching" field.

$$h_0 \sim \frac{|J_{\text{af}}| d_{\text{af}}}{d_{\text{f}} R^2} \propto R^{-2} \tag{6.25}$$

Unfortunately, we do not know the experimental observations of the predicted "switching" phenomenon. The transition of a layer (ferro- or antiferromagnetic) from a nanodomain to a homogeneous state can be observed by examining its surface using spin-polarization scanning tunneling microscopy.

6.3.5 Going Beyond the Exchange Approximation

Advances in nanostructure technologies make it possible to create nanostructures with R of the order of tens to hundreds of nanometers, which leads to the need to consider the phenomenon of "switching" beyond the framework of the exchange approximation, taking into account the power of single-ion anisotropy. That is what we will discuss below. If the characteristic distance between the edges of neighboring atomic steps at the interface R exceeds the thickness of the traditional domain wall in the antiferromagnet $\Delta_{\rm af}$, then the above "switching" scenario cannot be realized. Instead of a smooth transition, there is a first-order phase transition in a critical magnetic field. The power of the domain walls in the antiferromagnetic layer is higher than in the ferromagnetic layer. The value of the required magnetic field $h_{\rm cr}$ is found from the conditions of equality of the gain in the Zeeman power and the loss in the power of the domain walls.

Going beyond the exchange approximation becomes necessary when the domain wall width found in the exchange approximation [formula (6.23)] is compared with the value of $\Delta_{\rm af}$. The corresponding magnetic field h_0 can be found from the formula (6.25) by replacing R with $\Delta_{\rm af}$

$$h_0 \sim \frac{|J_{\rm af}|d_{\rm af}}{d_{\rm f}\Delta_{\rm af}^2}, \tag{6.26}$$

The view of the domain wall in the range of magnetic fields $h_{\rm cr} < h < h_0$ is shown in Fig. 6.5.

To estimate the characteristic width of the wall, we have limited ourselves to considering the case $d_{\rm af}|J_{\rm af}| > d_{\rm f}J_{\rm f}$. A simple power calculation taking into account the power of the single-ion anisotropy gives an estimate for the width of the domain wall

$$\tilde{\Delta} \approx \left(\frac{|J_{\rm af}|d_{\rm af}}{K_{\rm f}d_{\rm f} + K_{\rm af}d_{\rm af}}\right)^{1/2}, \tag{6.27}$$

where $K_{\rm f}$ and $K_{\rm af}$ are the uniaxial anisotropy constants of the ferro- and antiferromagnet. The refined value h_0 can be obtained by replacing in the formula (6.26) $\Delta_{\rm af}$ by $\tilde{\Delta}$:

$$h_0 \sim K_{\rm f} + K_{\rm af}d_{\rm af}/d_{\rm f}. \tag{6.28}$$

The main contribution to the power of the domain wall makes the part of the wall located in the antiferromagnet. Its structure and surface power density are similar to those of a traditional Bloch domain wall in an antiferromagnet [3]:

$$\tilde{\varepsilon} \approx b^{-2}[|J_{\rm af}|(K_{\rm af} + K_{\rm f}d_{\rm f}/d_{\rm af})]^{1/2}. \tag{6.29}$$

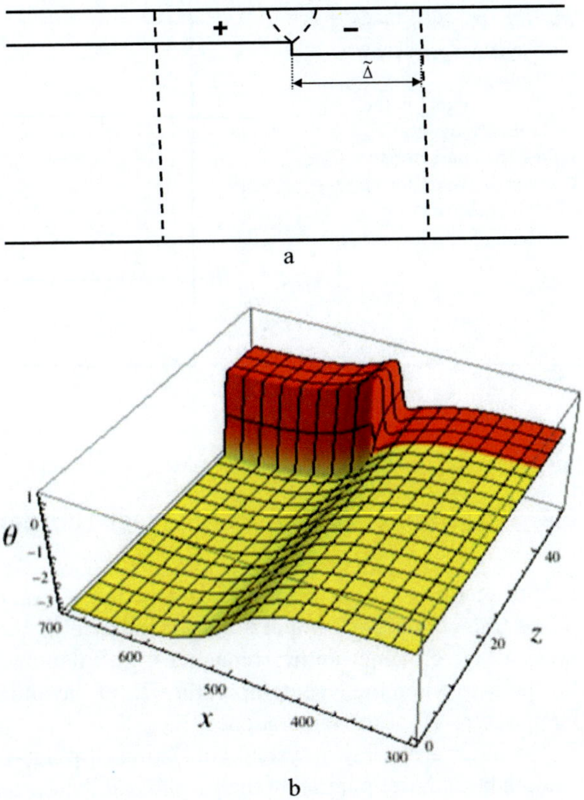

Fig. 6.5 The domain wall generated by the atomic step in the range of weak ($h = 0.0001$) magnetic fields: schematic representation (**a**) and the angle θ calculated for the case $J_{\mathrm{af}} = -1$, $J_{\mathrm{f}} = J_{\mathrm{f,af}} = 0.1$, $K_{\parallel} = 5 * 10^{-5}$ (**b**). The step at the interface between the ferromagnet ($z > 40$) and the antiferromagnet ($z < 40$) corresponds to the coordinate $x = 500$

The loss in the power of the domain wall due to "switching" per unit length of the edge of the atomic step is equal to $d_{\mathrm{af}}\varepsilon_{\mathrm{af}}$.

The gain in the power of interaction with the magnetic field due to "switching" in the domain bounded by the edges of two adjacent atomic steps is equal $2Rd_{\mathrm{f}}h/b^3$ to the unit length of the edge of the atomic step.

From the power balance, we get the value of the critical field h_{cr}

$$h_{\mathrm{cr}} = \frac{\tilde{\varepsilon} d_{\mathrm{af}} b^2}{R d_{\mathrm{f}}}. \tag{6.30}$$

Factor 2 disappears because there are two domain walls per one reorienting ferromagnetic domain.

In contrast to the domain of applicability of the exchange approximation, where $h_{\mathrm{cr}} \propto R^{-2}$, the critical field behaves as $h_{\mathrm{cr}} \propto R^{-1}$. Since there is no phase transition in the strict sense of the word in the field of applicability of the exchange approximation, a critical point C occurs in the phase diagram "magnetic field-characteristic distance between the edges of atomic steps at the interface" at $h \approx h_0$ (Fig. 6.6).

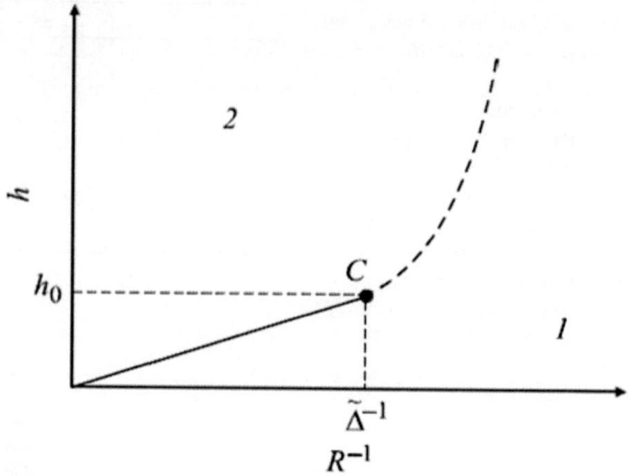

Fig. 6.6 Phase diagram of a two-layer nanostructure "ferromagnet-antiferromagnet" in the variables "magnetic field—the characteristic distance between the edges of atomic steps at the interface"

6.4 Exchange Bias Near the Neel Temperature

In this section, within the framework of the Landau phenomenological theory of phase transitions, the temperature dependence of the exchange bias value in the system "ferromagnet-antiferromagnet with uncompensated surface" is calculated for the case when the Neel temperature T_N of the antiferromagnet exceeds the Curie temperature T_C of the ferromagnet.

Consider a two-layer system consisting of plane-parallel layers with atomically smooth boundaries parallel to the xy plane of the Cartesian coordinate system. In this approximation, the system can be described in the framework of a one-dimensional model. Let us limit ourselves to the case of a single and common easy axis lying in the plane of the layers when the order parameters—the magnetization of the ferromagnet and the vector of antiferromagnetism can be considered as one component. The order parameters η_{FM} and η_{AFM} depend only on the z coordinate (the z-axis is perpendicular to the plane of the layers). Let the ferromagnet layer be located in the positive area of the z-axis, and the antiferromagnet layer in the negative area. Consider the case when the surface of an antiferromagnet is uncompensated; that is, it contains atoms of only one of the two sublattices of a collinear antiferromagnet. Then, the interaction power of the layers has the form $-J_{\text{INT}}\eta_{0\text{FM}}\eta_{0\text{AFM}}$, where $\eta_{0\text{FM}} = \eta_{\text{FM}}(0)$, $\eta_{0AFM} = \eta_{\text{AFM}}(0)$ and for certainty we put $J_{INT} > 0$.

In the course of calculations, we will assume that the thicknesses of the layers d_{FM}, d_{AFM} far exceed the radii of correlation of the order parameters corresponding to the layers. The expression for the free power per unit length along the z-axis in the framework of the Landau phenomenological theory of phase transitions is written as

$$\begin{aligned} F_{\text{FM}} &= F_{0\text{FM}} + A_{\text{FM}}(\eta_{\text{FM}}(z))^2 + B_{\text{FM}}(\eta_{\text{FM}}(z))^4 \\ &\quad + C_{\text{FM}}\left(\eta'_{\text{FM}}(z)\right)^2 - H\eta_{\text{FM}}(z) \\ F_{\text{AFM}} &= F_{0\text{AFM}} + A_{\text{AFM}}(\eta_{\text{AFM}}(z))^2 \end{aligned}$$

$$+ B_{AFM}(\eta_{AFM}(z))^4 + C_{AFM}\left(\eta'_{AFM}(z)\right)^2, \tag{6.31}$$

where A_{FM}, A_{AFM}, B_{FM}, B_{AFM}, C_{FM}, C_{AFM}, are the coefficients in the expansion of the free power by degrees of the order parameters and their derivatives, H is the external magnetic field, $A_{FM(AFM)} = A'_{FM(AFM)}\left(T - T_{C(N)}\right)$.

The total free power is defined as

$$\tilde{F} = \int_0^{d_{FM}} F_{FM}dz + \int_{-d_{AFM}}^{0} F_{AFM}dz + \Delta A_{FM}\eta_{0FM}^2 + \Delta A_{AFM}\eta_{0AFM}^2 - J_{INT}\eta_{0FM}\eta_{0AFM}, \tag{6.32}$$

where ΔA_{FM}, ΔA_{AFM} are the coefficients that take into account the change in local susceptibility near the interface of the layers. Here and in the future, we will measure the z coordinate in units of the interatomic distance along this axis. It is necessary to know the crystal structure of the system to determine ΔA_{FM}, ΔA_{AFM}. Let the ferromagnet and antiferromagnet have a volume-centered tetragonal lattice with an axis c lying in the plane of the layers (100), and the atoms located at its vertices are the closest to the atom located in the center of the cell. Then, at the interface, the four atoms closest to the ferromagnet atom are the atoms of the antiferromagnet; that is, they replace half of its nearest neighbors. Hence, in the nearest neighbor interaction approximation $\Delta A_{FM} = \frac{A'_{FM}T_C}{2}$, $\Delta A_{AFM} = \frac{A'_{AFM}T_N}{2}$.

The dependences of the order parameters on the coordinate $\eta(z)$ are determined by the solution of the Euler equations [4]

$$\eta''_{FM} = \alpha_{FM}\eta_{FM} + \beta_{FM}\eta_{FM}^3 - \gamma,$$
$$\eta''_{AFM} = \alpha_{AFM}\eta_{AFM} + \beta_{AFM}\eta_{AFM}^3 \tag{6.33}$$

where dimensionless parameters are entered

$$\alpha_{FM} = \frac{A_{FM}}{C_{FM}}, \Delta\alpha_{FM} = \frac{\Delta A_{FM}}{C_{FM}}, \beta_{FM} = \frac{2B_{FM}}{C_{FM}}, \gamma = \frac{H}{2C_{FM}}$$
$$\alpha_{AFM} = \frac{A_{AFM}}{C_{AFM}}, \Delta\alpha_{AFM} = \frac{\Delta A_{AFM}}{C_{AFM}}, \beta_{AFM} = \frac{2B_{AFM}}{C_{AFM}} \tag{6.34}$$

Assuming that at $z \to \infty$, $\eta'_{FM}(z) \to 0$, $\eta_{FM}(z) \to \eta_{\infty FM}$, and at $z \to -\infty$, $\eta'_{AFM}(z) \to 0$, $\eta_{AFM}(z) \to \eta_{\infty AFM}$, where $\eta_{\infty FM}$ and $\eta_{\infty AFM}$ are the equilibrium order parameters in a boundless sample, we find the first integrals of Eq. (6.33).

$$\frac{\left(\eta'_{FM}\right)^2}{2} = \frac{\alpha_{FM}\left(\eta_{FM}^2 - \eta_{\infty FM}^2\right)}{2} + \frac{\beta_{FM}\left(\eta_{FM}^4 - \eta_{\infty FM}^4\right)}{4} - \gamma(\eta_{FM} - \eta_{\infty FM})$$
$$\frac{\left(\eta'_{AFM}\right)^2}{2} = \frac{\alpha_{AFM}\left(\eta_{AFM}^2 - \eta_{\infty AFM}^2\right)}{2} + \frac{\beta_{AFM}\left(\eta_{AFM}^4 - \eta_{\infty AFM}^4\right)}{4}. \tag{6.35}$$

Solving Eq. (6.35) in quadratures, we find the dependences $\eta_{\mathrm{FM}}(z)$ and $\eta_{\mathrm{AFM}}(z)$ for the given order parameters at the interface of the system:

$$\int_{\eta_{0\mathrm{FM}}}^{\eta_{\mathrm{FM}}(z)} \left[\frac{\beta_{\mathrm{FM}}}{2}\left(\eta_{\mathrm{FM}}^4 - \eta_{\infty\mathrm{FM}}^4\right) + \alpha_{\mathrm{FM}}\left(\eta_{\mathrm{FM}}^2 - \eta_{\infty\mathrm{FM}}^2\right) - 2\gamma\left(\eta_{\mathrm{FM}} - \eta_{\infty\mathrm{FM}}\right)\right]^{-1/2} d\eta = z$$

$$\int_{\eta_{0\mathrm{AFM}}}^{\eta_{\mathrm{AFM}}(z)} \left[\frac{\beta_{\mathrm{AFM}}}{2}\left(\eta_{\mathrm{AFM}}^4 - \eta_{\infty\mathrm{AFM}}^4\right) + \alpha_{\mathrm{AFM}}\left(\eta_{\mathrm{AFM}}^2 - \eta_{\infty\mathrm{AFM}}^2\right)\right]^{-1/2} d\eta = z. \tag{6.36}$$

Having obtained the order parameters dependences numerically $\eta_{\mathrm{FM}}(z)$ and $\eta_{\mathrm{AFM}}(z)$ and substituting them into the integral (6.32), we obtain the dependence of the free power of the inhomogeneity on the order parameters at the interface $\tilde{F} = (\eta_{0\mathrm{FM}}, \eta_{0\mathrm{AFM}})$, from which we can obtain the values $\eta_{0\mathrm{FM}}^{ext}$ and $\eta_{0\mathrm{AFM}}^{ext}$ implementing its minimum.

For certainty, the equilibrium value of the order parameter in the antiferromagnet $\eta_{\infty\mathrm{AFM}}$ was chosen as positive. In this case, the hysteresis loop is shifted to the area of negative external fields by an amount H_{EB}.

At $T < T_{\mathrm{C}}$, that is, at $\alpha_{\mathrm{FM}} < 0$ and $\alpha_{\mathrm{AFM}} < 0$, there are three ranges of values of the external magnetic field: $(-\infty, -\gamma^*)$,$(-\gamma^*, \gamma^*)$ and $(\gamma^*, +\infty)$, where γ^* is given by the following expression

$$\gamma^* = \frac{2}{3}\left(\frac{|\alpha_{\mathrm{FM}}|^3}{3\beta_{\mathrm{FM}}}\right)^{1/2}. \tag{6.37}$$

In the first domain, there is real the only one, for certainty, the first root of the equation

$$\beta_{\mathrm{FM}}\eta_{\infty FM}^3 + \alpha_{\mathrm{FM}}\eta_{\infty FM} - \gamma = 0, \tag{6.38}$$

which determines the value of the order parameter far from the interface in the volume of the ferromagnet. In the second area all three roots are real and in the third area again the only one root is real, let it be the third one. To determine the value of the exchange bias in the field $(-\gamma^*, \gamma^*)$ area, the dependences of the free power of the system on the external field were numerically determined for a given value of the ferromagnetic order parameter in the volume equal, respectively, to the first and third roots of Eq. (6.38) (the solution corresponding to the second root is unstable). Then, the abscissa of the intersection point of the obtained curves was determined $\tilde{\gamma}$. If the condition $M_1(\tilde{\gamma}) \cdot M_3(\tilde{\gamma})$ is satisfied at this point, where $M_i(\tilde{\gamma})$ is a magnetic moment of the ferromagnet at a value $\eta_{\infty\mathrm{FM}}$ equal to the i-th root of the Eq. (6.38), then the field $\tilde{\gamma}$ will be the exchange bias field: $\gamma_{\mathrm{EB}} = \tilde{\gamma}$. If this condition is not met, then the exchange bias field will be the field at which the value $M_i(\tilde{\gamma})$ for i, which corresponds to the lower free power of the system, turns to zero.

For $T_C < T < T_{\mathrm{N}}$ ($\alpha_{\mathrm{FM}} > 0$ and $\alpha_{\mathrm{AFM}} < 0$) over the entire range of fields, only one of the roots of Eq. (6.38) is real. Therefore, the field of exchange bias corresponds to the field in which the magnetic moment of the ferromagnet vanishes.

At $T_N < T$, the exchange bias does not exist. But when the value of the integral of the exchange interaction between the layers at the interface exceeds the critical value J^*_{INT}, there is a surface phase transition (SPT) of the second order, as a result of which order parameters appear near the interface in the layers, whose values fall exponentially deeper into the layers.

Since the parameters η_{0FM}, η_{0AFM} turn to zero at the point of the SPT, we decompose the free power $\tilde{F}(\eta_{0FM}, \eta_{0AFM})$ in a series concerning these parameters and investigate the quadratic term in them $\tilde{F}_1$

$$\tilde{F}_1 = C_{FM}\big(\big(\sqrt{\alpha_{FM}} + \Delta\alpha_{FM}\big)\eta^2_{0FM} + \varsigma\big(\sqrt{\alpha_{AFM}} + \Delta\alpha_{AFM}\big)\eta^2_{0AFM} - \kappa\eta_{0FM}\eta_{0AFM}\big), \tag{6.39}$$

where $\varsigma = \frac{C_{AFM}}{C_{FM}}$, $\kappa = J_{INT}/C_{FM}$, $\alpha_{FM(AFM)} \equiv \alpha'_{FM(AFM)}\big(T - T_{C(N)}\big)$.

Above the SPT point, both eigenvalues of this quadratic form are positive, and at the transition point, one of the eigenvalues vanishes. This condition gives a parameter value κ that ensures the occurrence of SPT at a temperature of T_S:

$$\kappa_S = 2\left[\varsigma\left(\sqrt{\alpha'_{FM}(T_S - T_C)} + \frac{\alpha'_{FM}T_C}{2}\right)\left(\sqrt{\alpha'_{AFM}(T_S - T_C)} + \frac{\alpha'_{AFM}T_N}{2}\right)\right]^{1/2}. \tag{6.40}$$

The formula (6.40) sets the implicit dependence of the temperature of the SPT T_S on the system parameters. The critical value κ^* meets the condition $T_S = T_N T_S = T_N$

$$\kappa^* = \left[2\varsigma\alpha'_{AFM}T_N\left(\sqrt{\alpha'_{FM}(T_N - T_C)} + \frac{\alpha'_{FM}T_C}{2}\right)\right]^{1/2}. \tag{6.41}$$

In the future, we will limit ourselves to considering the case $\kappa < \kappa^*$ when there is no SPT.

Thus, the exchange shift field is determined as a function of the system parameters as a result of the calculations.

The simulation was performed for $T_C = 300\,\text{K}$, $T_N = 330\,\text{K}$, $d_{FM} = 200$. Following the microscopic calculation for the Ising model in the nearest neighbor interaction approximation [V. G. Wax. Introduction to the microscopic theory of ferroelectrics. Nauka, M. (1973). 328 p], the following values of the coefficients of the expansion of free power by degrees of the order parameter were selected [5]:

$$\varsigma = T_N/T_C = 1.1;\ \alpha_{FM} = (T - T_C)/T_C;\ \alpha_{AFM} = (T - T_N)/T_N;$$
$$\beta_{FM} = \beta_{AFM} = 1/3$$

The dependence of the value of the exchange shift γ_{EB} on the temperature for different values κ is shown in Fig. 6.7.

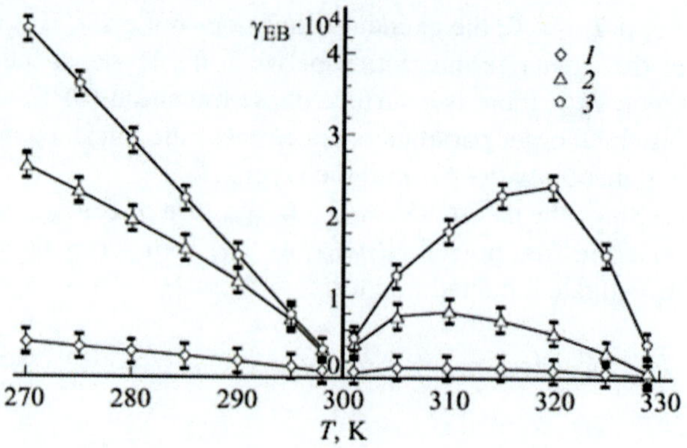

Fig. 6.7 Temperature dependence of the exchange bias field: rhombus—$\kappa = 0.1\kappa^*$, triangle—$\kappa = 0.5\kappa^*$, pentagon—$\kappa = 0.9\kappa^*$

Near the Curie temperature, our assumption that the thickness of the ferromagnet far exceeds the correlation radius of the order parameter is violated r_c, so the calculation cannot be performed in this area. A similar situation occurs in an antiferromagnet near the Neel temperature.

In the area of the Curie temperature, the minimum value of the exchange bias is observed. A monotonous decrease H_{EB} with increasing temperature in the area $T < T_C$ was predicted in [X. Zhi-Jie, W. Huai-Yu, D. Ze-Jun. Chinese Physics **16**, 2123 (2007)] based on Green's function method for the case $T_N \approx T_C$.

It is possible to explain the course of the curve qualitatively, proceeding from the following considerations. The value of the exchange bias can be estimated by equating the interaction power of the layers to the Zeeman power of a ferromagnet in an external magnetic field:

$$H_{EB} = \frac{E_{INT}}{\mu_0 M d_{FM}}, \tag{6.42}$$

where E_{INT} is the surface power density of the interaction of the layers, M is the magnetization of the ferromagnet, and μ_0 a is the SI constant. Near the Curie temperature, the value η_{0FM} is determined by the action of the exchange field of the antiferromagnet, and therefore, E_{INT} practically does not change at $T \to T_C$. The magnetization of a ferromagnet in a linear field approximation consists of the spontaneous magnetization of M_0 and the field-induced summand χH. Due to the inequality $d_{FM} \gg r_c$, the area's contribution adjacent to the interface of the layers to the magnetization can be neglected. When $H = H_{EB}$ from Eq. (6.42), we obtain the following quadratic equation for determining the value of the exchange bias:

$$\mu_0 \chi d_{FM} H_{EB}^2 + \mu_0 M_0 d_{FM} H_{EB} - E_{INT} = 0. \tag{6.43}$$

Since at $T \to T_C$, $M_0 \to 0$, and the susceptibility of the ferromagnet χ diverges, then near T_C in the mean-field approximation

$$H_{EB} = \left(\frac{E_{INT}}{\mu_0 \chi d_{FM}} \right)^{1/2} \propto |T - T_C|^{1/2}. \tag{6.44}$$

Of course, due to the non-linearity of the magnetic susceptibility, the value of the exchange bias at $T \to T_C$ is not exactly equal to zero.

Since the exchange field created by the antiferromagnet at the interface at $T \to T_N$ is proportional to $\eta_{\infty AFM}$, and the ferromagnet's susceptibility has no specific features at the Neel temperature, then at $T \to T_N$

$$H_{EB} \propto \eta_{\infty AFM} \propto |T - T_N|^{1/2}.$$

Consequently, there is a maximum exchange bias between the temperatures T_C and T_N, which is confirmed by the experimental data [X. W. Wu, C. L. Chien. Phys. Rev. Lett. **81**, 2795 (1998)].

Thus, the existence of an exchange bias in the temperature $T_C < T < T_N$ range is due to the appearance of an order parameter in a ferromagnet induced by its interaction with an antiferromagnet.

The presence of a minimum value of the exchange bias near the Curie temperature is due to an increase in the magnetic susceptibility of the ferromagnet.

References

1. Levchenko VD, Morosov AI, Sigov AS (2002) Unidirectional anisotropy and roughness of the ferromagnetic-antiferromagnetic interface. FTT 44(1):128–134
2. Morosov AI, Sigov AS (2002) Unidirectional anisotropy in the ferromagnetic-antiferromagnetic system. FTT 44(11):2004–2009
3. Morosov AI, Sigov AS (2002) Interface roughness and unidirectional anisotropy of thin ferromagnetic film on the uncompensated surface of the antiferromagnet. J Magn Magn Mater 242–245(P2):1014–1016
4. Morosov AI, Markets DO (2011) Exchange shift in a two-layer ferromagnet-antiferromagnet system with close phase transition temperatures. FTT 53(1):66–69
5. Berzin AA, Morosov AI, Sigov AS (2012) "Switching" of the nanodomain state of the frustrated ferromagnet-antiferromagnet system by an external magnetic field. FTT 54(11):2090–2094
6. Berzin AA, Morosov AI, Sigov AS (2013) The phase transition of the "switching" of the nanodomain state of the frustrated ferromagnet-antiferromagnet system. FTT 55(4):702–704

Chapter 7
Spin-Valve Structure Ferromagnet–Antiferromagnet–Ferromagnet

7.1 Frustrations in a Three-Layer System

Consider a spin-valve structure consisting of two ferromagnetic layers separated by an antiferromagnetic layer.

If the atomic planes parallel to the interface are uncompensated, then nuclear steps at the interface lead to frustration of the exchange interaction between the ferromagnetic layers. Indeed, on one side of the edge of the atomic step, the number of atomic planes in the antiferromagnet layer is even, and on the other it is odd. In the case of an odd number of uncompensated planes of the antiferromagnet, the spins of the ferromagnetic layers interact with the nearest spins of the antiferromagnet belonging to the same sublattice (Fig. 7.1a). For any sign of the exchange integral $J_{\mathrm{f,af}}$ between adjacent spins located in different layers, the parallel orientation of the magnetizations of the ferromagnetic layers is energetically advantageous. Suppose the number of atomic planes in the antiferromagnetic layer is even. In that case, the spins of the ferromagnetic layers interact with the spins of the antiferromagnet closest to them, which belong to different sublattices. The antiparallel orientation of the magnetizations of the ferromagnetic layers is energetically advantageous (Fig. 7.1b); the atomic step generates frustration.

Thus, the atomic steps at both interfaces divide the plane parallel to the layers into two types of areas: in the areas of the first type, the parallel orientation of the magnetizations of the ferromagnetic layers is energetically advantageous, and in the areas of the second type, the antiparallel direction.

The exchange approximation corresponds to the case when the width of the atomic steps R and the thickness of the layers are much smaller than the width of the traditional domain walls in ferro- and antiferromagnets, determined by the ratio of the exchange power and the anisotropy power. The achievements of modern technology make it possible to obtain films and multilayer structures with a large width of atomic steps at the interfaces, which makes it essential to go beyond the exchange approximation.

A. S. Sigov, *Multilayer Magnetic Nanostructures*, Springer Aerospace Technology,
https://doi.org/10.1007/978-981-19-6246-2_7

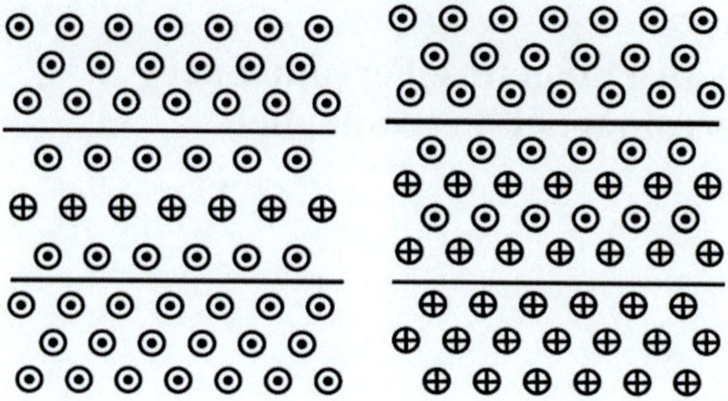

Fig. 7.1 Orientation of the spins corresponding to the minimum power, **a** for an odd, **b** for an even number of atomic planes in the antiferromagnetic layer

In this chapter, we will limit ourselves to considering the case $\gamma_{\mathrm{af}} \ll 1$. Otherwise, there are practically no distortions of the order parameter in the antiferromagnetic layer. The problem is reduced to distortions of the order parameter in a ferromagnetic film on a rigid antiferromagnetic substrate, discussed in the previous chapters.

In addition, to reduce the number of parameters, we will assume that the layer thicknesses and the single-ion anisotropy constants are the same.

7.2 Domain Walls in a Three-Layer System

The most important geometric parameters of the system are the layer thickness a and the characteristic distance between the edges of the steps at the boundaries of the layers R. If the value R exceeds a specific critical value, the value of which we will specify later, then in the exchange approximation, all layers are divided into domains with parallel and antiparallel orientation of the magnetizations of the ferromagnetic layers. The domain wall of a new type, generated by frustration, permeates all three layers; its coordinates in the plane of the layer coincide with the edge of the atomic step on any of the two interfaces. The turn of the magnetizations of the ferromagnetic layers in the domain wall in the exchange approximation occurs in opposite directions at an angle of 90°. The antiferromagnetic order parameter turns together with the magnetization of the ferromagnetic layer whose interface at this point does not contain a step (Fig. 7.2).

The structure and power of the domain wall depend on the parameter $\gamma_{\mathrm{f,\,af}}a/\gamma_{\mathrm{af}}$.

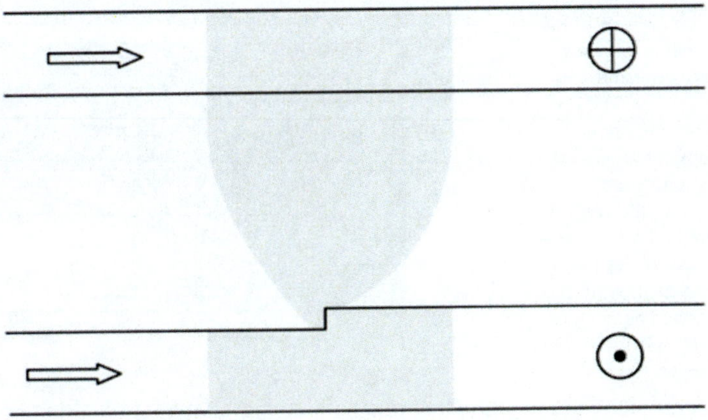

Fig. 7.2 Frustration-induced distribution of the order parameters near the domain wall, calculated in the exchange approximation

7.2.1 $\gamma_{f,\,af}a/\gamma_{af} \ll 1$

In this case, the dependence $\theta_{\mathrm{f,af}}(z)$, that is, the broadening of the domain wall can be ignored, and the problem becomes one-dimensional. The domain wall takes the form shown in Fig. 7.3a and is characterized by two widths δ_{f} and δ_{af} corresponding to the ferromagnetic and antiferromagnetic layers.

The inhomogeneity power of the ferromagnetic order parameter in one layer, per unit length of the domain wall, is given by the formula (4.16). Similarly, the inhomogeneity power of the antiferromagnetic order parameter is [1]

$$w_{\mathrm{af}} = \frac{|J_{\mathrm{af}}|a}{b\delta_{\mathrm{af}}}. \tag{7.1}$$

Let the atomic step be located at the first interface of the layers, and it corresponds to the coordinate $x = 0$.

The interaction power of the layers at this boundary per unit length of the domain wall is given by the formula

$$w_{f1,af} = \frac{J_{\mathrm{f,\,af}}}{b}\left\{\int\limits_{-\infty}^{0}\left[1-\cos\left(\theta_{f,1}(x)-\theta_{\mathrm{af}}(x)\right)\right]\mathrm{d}x + \int\limits_{0}^{+\infty}\left[1+\cos\left(\theta_{f,1}(x)-\theta(x)\right)\right]\mathrm{d}x\right\} \tag{7.2}$$

At the second interface, where there is no step at this point, the same power is equal to

$$w_{\mathrm{f2,\,af}} = \frac{J_{\mathrm{f,\,af}}}{b}\int\limits_{-\infty}^{+\infty}\left[1-\cos\left(\theta_{\mathrm{f,\,2}}(x)-\theta_{\mathrm{af}}(x)\right)\right]\mathrm{d}x. \tag{7.3}$$

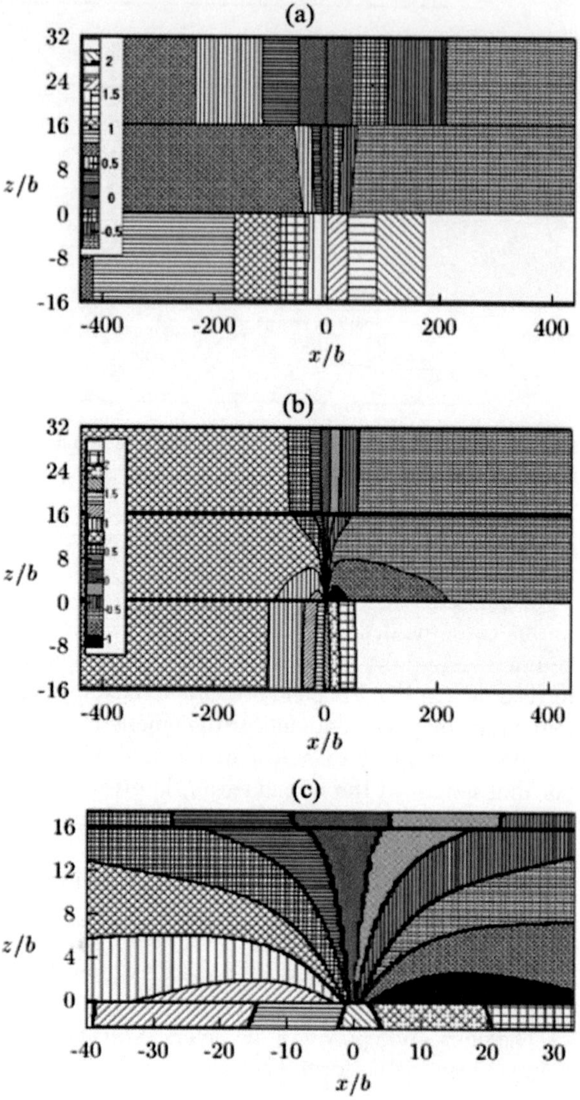

Fig. 7.3 Domain wall in the spin-valve system in cases $\gamma_{\mathrm{f,\,af}}a/\gamma_{\mathrm{af}} \ll 1$ (**a**) and $\gamma_{\mathrm{f,\,af}}a/\gamma_{\mathrm{af}} \gg 1$ (**b**). The box shows the correspondence between the hatching and the value $\theta_{f(af)}$ measured in radians. **c** Shows the central part of figure **b**. The coordinates $z = 0$ and $z = 16$ correspond to the boundaries of the partition. The edge of the atomic stage is located at the point $x = z = 0$

To estimate the parameters δ_{f} and δ_{af} use the wall symmetry, that is, that in the first layer $\theta_{\mathrm{f,\,1}}(0) = -\pi/4$, and $\theta_{\mathrm{af}}(0) = \theta_{\mathrm{f,\,2}}(0) = \pi/4$, with an accuracy of order corrections γ_{af}. In addition, we will use the fact that as will be shown below, $\delta_{\mathrm{af}} \ll \delta_{\mathrm{f}}$. Then the area of integration over x can be divided into the area $|x| \le \delta_{\mathrm{af}}$ in which $\theta_{\mathrm{f},i}(x)$ it can be considered equal to $\theta_{\mathrm{f},i}(0)$ $(i = 1, 2)$, and the area $\delta_{\mathrm{af}} \le |x| \le \delta_{\mathrm{f}}$ in which the value θ_{af} can be regarded as close to 0 or $\pi/2$, respectively, to the left and right of the edge of the step. Then

$$w_{f2,\,af} + w_{f1,\,af} \sim \frac{J_{f,\,af}}{b}(c_1\delta_f + c_2\delta_{af}), \tag{7.4}$$

where the coefficients c_1 and c_2 are positive, and their magnitude is of the order of unity.

Minimizing the total power

$$w = w_{f,\,1} + w_{af} + w_{f,\,2} + w_{f1,\,af} + w_{f2,\,af}$$

gives in order of magnitude

$$\delta_f \approx \left(a/\gamma_{f,af}\right)^{1/2}, \tag{7.5}$$

$$\delta_{af} \approx \left(\frac{a\gamma_{af}}{\gamma_{f,\,af}}\right)^{1/2} = \delta_f\gamma_{af}^{1/2} \ll \delta_f, \tag{7.6}$$

which justifies the assumption.

The total power of the domain wall per unit length in the plane of the layers is on the order of magnitude equal to

$$w \sim \frac{J_f}{b}\left(a\gamma_{f,af}\right)^{1/2}. \tag{7.7}$$

Numerical calculations in a wide range of parameters confirm the estimates made.

7.2.2 $\gamma_{f,\,af}a/\gamma_{af} \gg 1$

In this more relevant case, the width of the domain wall in the antiferromagnetic layer significantly increases as it moves away from the interface containing the atomic step, as in the case $\gamma_{af} \ll 1$ of a two-layer system (Fig. 7.3b). Its minimum thickness δ_0 is found similarly and is given by the formula (4.14).

The thickness of the domain wall in ferromagnetic layers δ_f can be obtained from simple power considerations, as in the previous section.

Suppose that $\delta_f \gg a$, and the step is located at point $x = 0$. Then, in the area $a \ll |x| \ll \delta_f$, the lines of the constant value θ_{af} are almost parallel to the layer interfaces (Fig. 7.3c). In this area $|\nabla\theta_{af}| \approx a^{-1}$. Then the distortion power of the antiferromagnetic order parameter per unit length of the domain wall is [2]

$$w_{af} = \frac{|J_{af}|}{b}\left(\frac{\delta_f}{a} + \ln\frac{a}{\delta_0}\right), \tag{7.8}$$

and w_f is given by the formula (4.16).

Minimizing the total power of the wall, we obtain

$$\delta_f \approx a/\gamma_{af}^{1/2} \gg a, \tag{7.9}$$

as it was assumed, therefore, its broadening in the ferromagnetic layer can be neglected.

Let us consider how the domain wall structure will change regarding the power of the single-ion anisotropy. The condition for the applicability of the exchange approximation a, $R \ll \min(\Delta_f, \Delta_{af})$ can be violated in two ways: by increasing the width of the steps for the case of thin layers ($a \ll \Delta_{af}$) and increasing both the width of the steps and the layer thickness.

If two mutually perpendicular easy axes lie in the plane of the layer, the pattern of partition of thin layers will not change. However, if there is only one easy axis in the plane of the layer, in the case $R \gg \Delta_f$ of partitioning thin layers into 90° domains, it ceases to be energetically advantageous. There is a situation similar to the one discussed in the third chapter. Indeed, let the magnetizations in the domains of one type be directed parallel to the "easy" axis, then in the domains of the second type, they will be perpendicular to it. There is an increase in the anisotropy power. The minimum of the total power is realized when both in domains with the parallel orientation of the magnetizations of the ferromagnetic layers and domains with antiparallel orientation of the magnetizations, the magnetizations in the domains are collinear to the "easy" axis. The structure of the domain walls perpendicular to the layers and separating these domains changes dramatically: they now consist of three areas, if we move along the coordinate perpendicular to the plane of the domain wall separating the domain with the parallel orientation.of the magnetizations of the ferromagnetic layers from the domain with the antiparallel orientation of the magnetizations. In the first area, both magnetization vectors are rotated in the plane of the layer at an angle of 45° from the "easy" axis. The width of this area is about the width of the traditional domain wall Δ_f, and with this turn, the magnetizations of the ferromagnetic layers remain parallel to each other.

The second area is the domain wall described above, generated by frustration. In it, the magnetization of one ferromagnetic layer rotates 90 degrees, preserving the direction of rotation in the first area, and the magnetization of the second layer rotates 90º in the opposite direction. As a result, the magnetizations will turn from the original direction at angles of 135º and -45° respectively (Fig. 3.2), and their orientation will become antiparallel.

In the last, third, area, the magnetizations, while remaining antiparallel, turn at an angle of 45° in the same direction as in the first area. At the exit from the wall, they form 180° and 0° angles with an "easy" axis, respectively. The antiferromagnetic order parameter turns together with the magnetization of the second ferromagnetic layer, the boundary with which at this point does not contain a step (Fig. 7.2).

With increasing layer thicknesses, the estimate (7.9) remains valid until the δ_f wall width generated by the exchange interaction frustration is compared in order of magnitude with the width of the traditional domain wall. After that, due to the contribution of the anisotropy power, the wall thickness reaches a constant value equal

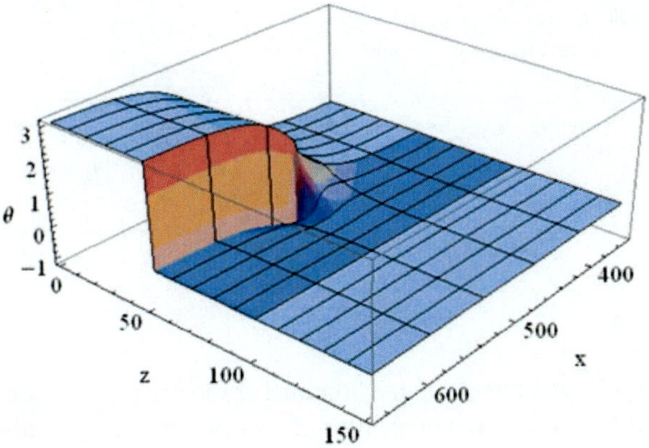

Fig. 7.4 A domain wall permeating a single layer, for $\gamma_{af} = 0.1$, $a = 50$, and $\Delta_f = 14$. The edge of the atomic step is located at $x = 500$, $z = 50$

to Δ_f. The condition of applicability of the exchange approximation $a\gamma^{1/2} \ll \Delta_f$ is equivalent under our assumptions to the condition $a \ll \Delta_{af}$.

As the thickness of the layers increases, at $a \sim \Delta_{af}$, the transformation of the domain walls consisting of three areas perpendicular to the layers occurs. The width of the central area of the wall is compared with the widths of the other two areas, and the domain wall in the second ferromagnetic layer disappears. The domain boundary in the first ferromagnetic layer, at the boundary with which the atomic step is present at this point, is transformed into the traditional 180° domain wall (Fig. 7.4).

Distortions of the antiferromagnetic order parameter capture the area around the edge of the atomic step, the size of which in the direction parallel to the layers (but perpendicular to the edge of the step), is of the order of Δ_f, and in the direction perpendicular to the layers, it is of the order of Δ_{af}. Consequently, the central part of the antiferromagnetic layer is in a monodomain state. The view of the solitary domain wall generated by the frustration of the exchange interaction at $a \gg \Delta_{af}$ is shown in Fig. 7.4.

7.3 Phase Diagram

In the exchange approximation, the polydomain phase is realized at $R > \delta_f$. This condition is met for $a \sim 1$ nm and $\gamma_{af} \sim 0.1$ already at a domain size of about 10 nm. Thus, we can talk about the nanodomain state, the study of which requires subtle techniques.

In the work [A. Paul. *J. Magn. Magn. Mater.* **240**, 497 (2002)], a multilayer Fe/Cr structure was studied in which the average thickness of the antiferromagnetic layers corresponded to the antiparallel orientation of the magnetizations of the neighboring ferromagnetic layers. It was found that with increasing roughness of the interlayer boundaries, the proportion of areas with the parallel orientation of the magnetizations of neighboring ferromagnetic layers increases and reaches 50%.

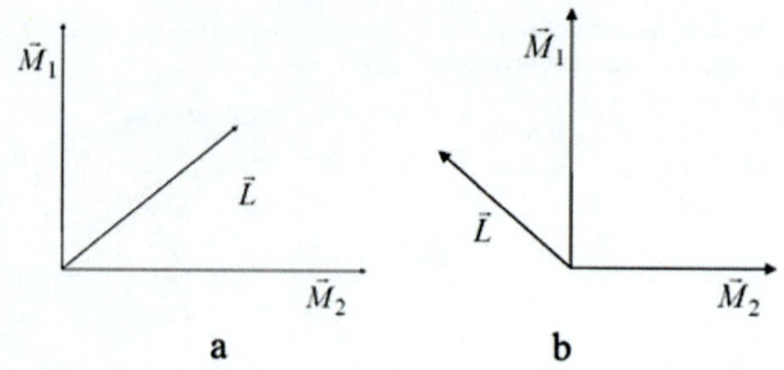

Fig. 7.5 Orientation of order parameters in domains

What happens to the domains as the parameter decreases R? In the field of applicability of the exchange approximation, the domain walls begin to overlap. At a critical value $R_c \sim \delta_f$, a continuous transition occurs to a state in which the ferromagnetic layers are almost homogeneous, with only weak magnetization distortions occurring in them. Since the Neel temperature of the interlayer TN (TN is less than the Curie temperature of the ferromagnet TC) parameter $\gamma_{af} \infty T_N - T$, this transition can be realized by heating the system from the initial temperature $T_0 < T_N$.

When $\gamma_{f,\,af} a/\gamma_{af} \ll 1$ in the range of values $\delta_{af} \ll R \ll \delta_f$, the antiferromagnetic layer is divided by the domain walls into 90° domains. The vector of antiferromagnetism in the domains of the first type looks along the bisector of the angle formed by the magnetizations (Fig. 7.5a), and in the domains of the second type it is orthogonal to this bisector (Fig. 7.5b). The type of area determines the type of domain, that is, the number (even or odd) of atomic planes in the antiferromagnetic layer.

The additional (relative to the state without frustrations) power is related to the interaction power of the layers and has the form [3]

$$W = -\frac{2J_{f,\,af}}{b^2}\left(\sigma_1 \cos\frac{\Psi}{2} + \sigma_2 \cos\frac{\pi - \psi}{2}\right), \tag{7.10}$$

where σ_1 and σ_2 are the areas of the first and second type regions on the surface of the antiferromagnetic layer, and ψ *is* the angle between the magnetizations of the ferromagnetic layers. For $\sigma_1 = \sigma_2$ the minimum power is achieved at $\psi = \pi/2$, that is, in the absence of an external magnetic field, there is a mutually perpendicular orientation of the magnetizations of the ferromagnetic layers.

If $a \ll R \ll \delta_{af}$, then the system is located in the area of weak distortions of the magnetic order parameters, and their orientation corresponds to the one shown in Fig. 7.5 a.

If $\gamma_{f,\,af} a/\gamma_{af} \gg 1$ the range of values $a \ll R \ll \delta_f$, the additional (relative to the state without frustrations) power is associated with the power of the two types of exchange spirals that occur in the interlayer. Each type of spiral is implemented in an area of the corresponding type. This state is described by the "magnetic proximity" model by Slonczewski [J. C. Slonczewski *J. Magn. Magn. Mater.* **150**, 13 (1995)]; namely, the dependence of the power of the system on the angle ψ between the magnetizations of the ferromagnetic layers is described by the formula (4.20), where

ψ and $\pi - \psi$ are the rotation angles of the antiferromagnetic order parameter in the exchange spirals of the first and second types, respectively. The estimate for the constants C_1 and C_2 gives:

$$C_{1,2} \approx \frac{J_{\mathrm{af}}}{a} * \frac{\sigma_{1,2}}{b^2}. \tag{7.11}$$

For $\sigma_1 = \sigma_2$ the minimum power is achieved at $\psi = \pi/2$, in the absence of an external magnetic field, there is a mutually perpendicular orientation of the magnetizations of the ferromagnetic layers. The condition that the power of the exchange spirals far exceeds the power of the anisotropy gives us a familiar inequality $a \ll \Delta_{\mathrm{af}}$. The magnetization of the ferromagnetic layers will be positioned relative to the "easy" axis so that it is the bisector of the right angle between them to reduce the anisotropy power.

The non-collinear orientation of the magnetizations of ferromagnetic layers in multilayer structures with an antiferromagnetic layer has been discussed repeatedly [see, for example, A. Shreyer et al. *Phys. Rev. B* **52**, 16,066 (1995); V. V. Ustinov et al. *JETF* **109**, 477 (1996)]. According to the data of neutron diffraction experiments [A. Schreyer et al. *Phys. Rev. Lett.* **79**, 4914 (1997); P. Bodeker, A. Schreyer, H. Zabel. *Phys. Rev. B* **59**, 9408 (1999)], an example of ferromagnetic–uncompensated antiferromagnetic structures are Fe/Cr structures. At a thickness *of* $a < 45$ Å, the chromium layers are a set of ferromagnetic planes with antiparallel spin orientation in adjacent planes. The spins of the chromium atoms lie in these planes, which, in turn, are parallel (on average) to the boundaries of the layers. A similar magnetic structure was observed in manganese layers in multilayer Fe/Mn structures [M. Chirita et al. *Phys. Rev. B* **58**, 869 (1998); S. Yan et al. *Phys. Rev. B* **59**, R11641 (1999)]. Thus, the above theory is applicable to the Fe/Cr and Fe/Mn structures, and it is on these structures that a series of experiments to verify it is possible.

However, it should be noted that in the absence of proper spatial resolution, it is difficult to distinguish between the polydomain state of ferromagnetic layers with nanometer-scale domains and the non-collinear form of homogeneous ferromagnetic layers only based on data averaged over the surface of the layer. In most cases, the authors of experimental works do not consider the possibility of the existence of a nanodomain state at all.

Let us now consider the range of values $R \ll a$. In this case, all the distortions are concentrated near the boundaries of the layers, the interaction between the ferromagnetic layers becomes weak, and the central role is played by the power of interaction between neighboring layers, discussed in Chap. 4.

As a result, for $\sigma_1 = \sigma_2$, antiferromagnetic order parameter is oriented perpendicular to the magnetizations of the ferromagnetic layers, which thus turn out to be collinear.

The simulation shows that the transition from a phase with mutually perpendicular magnetizations of ferromagnetic layers to a stage with their collinear orientation is the first order phase transition. Both states coexist in the whole range of R values, and their energies are compared at some value $R^* \sim a$. The value of R^* does not

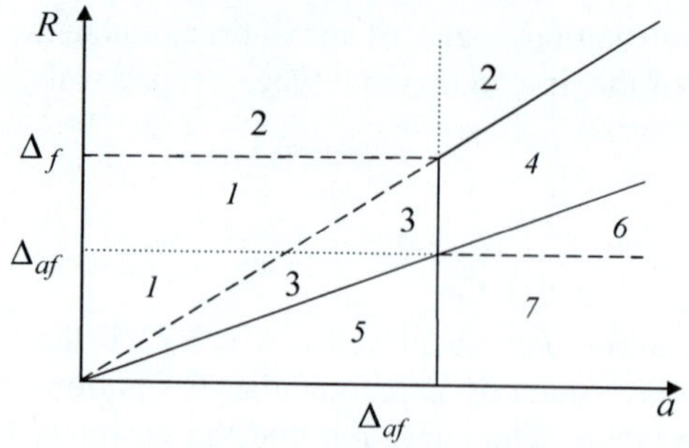

Fig. 7.6 Phase diagram of the spin-valve system: 1—phase with 90° domains in the ferromagnetic layers, 2—phase with 180° domains in the ferromagnetic layers, 3 - phase with single-domain ferromagnetic layers and the exchange spirals in antiferromagnetic interlayer (Slonczewski), 4—phase with 180° domains in the antiferromagnetic layer, 5—phase static 90° spin vortices and antiferromagnetic order parameter perpendicular to magnetization of ferromagnetic layers, 6—phase with 180° exchange spirals parallel to interfaces, in antiferromagnetic layer 7—phase with spin vortices and two 90° exchange spirals, parallel to interfaces, in the antiferromagnetic layer. Solid lines correspond to the lines of first order phase transitions

depend on temperature, so this phase transition cannot be observed by changing the system's temperature [4].

The phase diagram "layer thickness—the characteristic distance between the edges of the atomic steps" of the spin-valve system for the case $J_{f,af} \sim J_{\mathrm{af}}$ when there is no area of weak distortion is shown in Fig. 7.6.

Let us consider how phase 3 changes in the thickness area $a > \Delta_{\mathrm{af}}$ when the power of the single-ion anisotropy becomes significant, for the case of a single easy axis in the plane of the layers. The exchange spirals in the antiferromagnetic layer turn into 180° domain walls of the following type: at a distance of an order of Δ_{af} from the edges of the atomic steps bounding the domain wall, the width of the domain wall increases from the interatomic distance to a value of an order of magnitude Δ_{af}, and in the rest of the area the wall is a traditional domain wall.

Thus, only the antiferromagnetic layer is now in the polydomain state (phase 4 in Fig. 7.6). The magnetizations of the ferromagnetic layers are collinear, and the transition from their parallel to antiparallel orientation should be accompanied by a collective "switching" of the domain walls in the antiferromagnetic layer (see Fig. 7.7). Due to the random location of the edges of the atomic steps, the energies of the initial and "switched" states are the same.

The transition from the monodomain state of ferromagnetic layers to the polydomain state (phase 2 in Fig. 7.6) in this thickness range is a first order phase transition.

In the case ($a \gg \Delta_{\mathrm{af}}$), when the transition area near the edge of the step gives a small contribution to the power of the domain wall, the value R_c corresponding to the phase transition point at a given layer thickness can be easily found from the following considerations. The power of the transverse domain walls in the polydomain phase per unit area of the layers is equal to

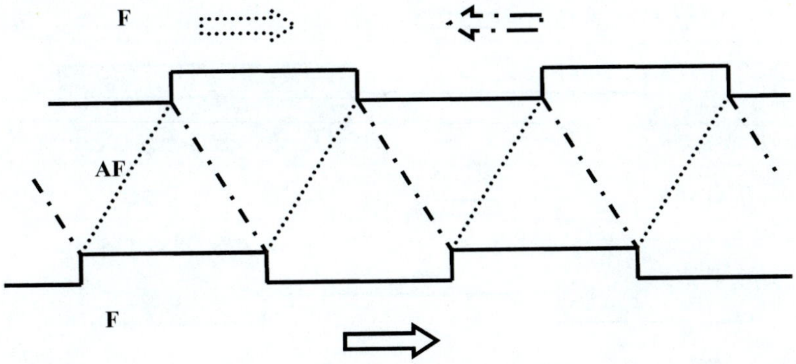

Fig. 7.7 Switching of the domain walls in the interlayer (**a**). Dotted-dashed and dotted lines show the positions of the domain walls in the antiferromagnetic layer before and after switching. Dotted-dashed and dotted arrows show the direction of magnetization of the upper layer before and after switching. Respectively, a solid arrow shows the direction of magnetization of the lower layer

$$w_p \sim \varepsilon_f a/R. \tag{7.12}$$

The power of the domain walls in the antiferromagnetic layer in the monodomain phase per unit area of the layer is equal to

$$w_m \sim \varepsilon_{af} L/R, \tag{7.13}$$

where L is the characteristic distance between the nearest steps at opposite interfaces (Fig. 7.7). Equating these energies and estimating L as $\left(a^2 + R^2/4\right)^{1/2}$, we find

$$R_c \sim a/\gamma_{af}^{1/2}. \tag{7.14}$$

At $R > R_c$, the ferromagnetic layers are in a polydomain state, and at $R < R_c$, in a monodomain state. In a natural spin-valve system, the phase transition is blurred due to the spread over the widths of the atomic steps at the interfaces.

It is easy to understand, looking at Fig. 7.7, that it is energetically advantageous for the domain boundary in an antiferromagnet to connect the steps at opposite layer boundaries only if the parameter R exceeds a particular critical value R^*. For the ordered arrangement of the steps shown in Fig. 7.7, it can be estimated as

$$R^* = (4/3)^{1/2} a. \tag{7.15}$$

At $R < R^*$, the distortions of the antiferromagnetic order parameter are concentrated near the boundaries of the layers, and the volume of the antiferromagnet is in the monodomain state (phase 6 in Fig. 7.6). The interaction between the ferromagnetic layers becomes weak, and the central role is played by the power of interaction between neighboring layers. The transition of an antiferromagnet from a polydomain

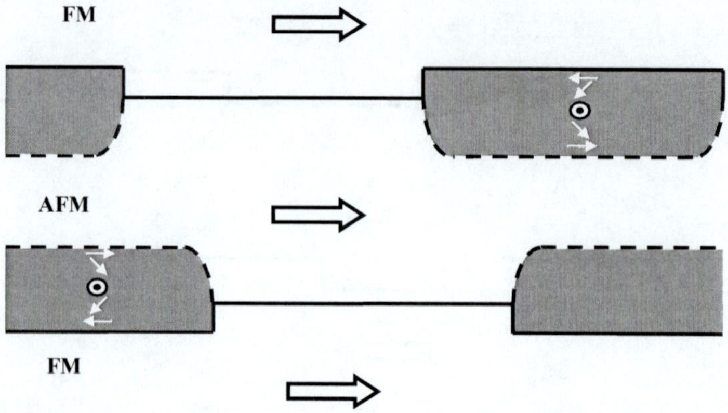

Fig. 7.8 Exchange spirals parallel to the interfaces in the antiferromagnetic layer. The arrows indicate the direction of the corresponding order parameter. The rotation of the antiferromagnetic order parameter occurs in the plane of the layers

to a monodomain state is also a first order phase transition of the, smeared due to the spread over the widths of the atomic steps at the interfaces.

At $\Delta_{af} \ll R < R^*$, the distortions of the antiferromagnetic order parameter have the form of 180° exchange spirals parallel to the interface, which exist in those areas of the interface where the spins of the ferromagnet are adjacent to the spins of one of the sublattices of the antiferromagnet (for certainty, the second one). In the areas where the spins of the ferromagnet are adjacent to the spins of the first sublattice of the antiferromagnet, the antiferromagnetic order parameter remains uniform (Fig. 7.8).

As the parameter decreases R in the range of values $R \ll \Delta_{af} \ll a$ in the interlayer, static spin vortices arise in both types of areas, whose edges at the interface coincide with the edges of the steps, and the size in the direction perpendicular to the boundary is the same as the minimum size of the vortex in the plane of the layers interface (Fig. 7.9).

To reduce the power of the vortices, the antiferromagnetic order parameter near the interface of the layers is oriented perpendicular to the "easy" axis. In this case, the rotation of the antiferromagnetism vector in each vortex occurs at an angle of 90°, differing in the direction of rotation in different types of areas. Along with static 90° vortices, 90° exchange spirals are formed in the interlayer near both interfaces so that the order parameter in the center of the antiferromagnetic layer is parallel to the "easy" axis (Fig. 7.9 a)—phase 7 in Fig. 7.6.

With a decrease in the parameter a at $a \sim \Delta_{af}$, these domain walls disappear abruptly, and the spin-valve system passes into a state with static 90° spin vortices and an antiferromagnetic order parameter oriented perpendicular to the magnetizations of the ferromagnetic layers and the "easy" axis (Fig. 7.9b)—phase 5 in Fig. 7.6.

In the presence of two mutually perpendicular easy axes lying in the plane of the layer, the phase diagram (Fig. 7.6) is modified (Fig. 7.10) Domain walls consisting of three areas are not formed, and the entire area of existence of the polydomain

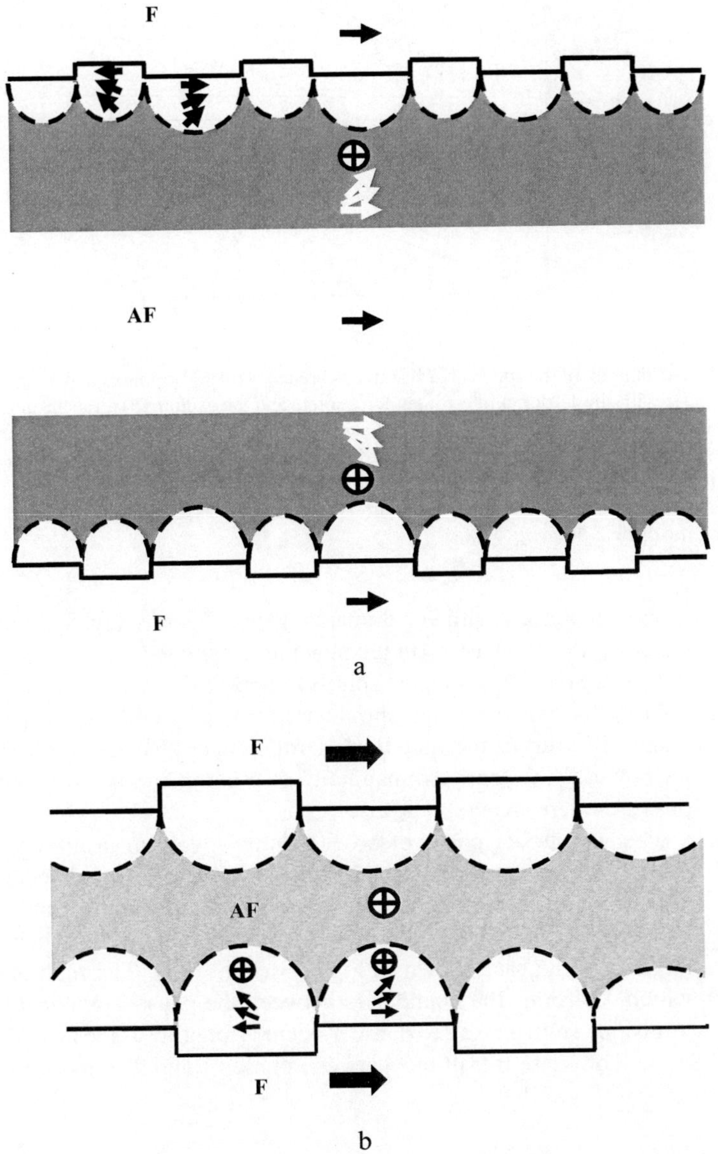

Fig. 7.9 Spin vortices and two 90° exchange spirals in an antiferromagnetic layer (**a**), a phase with spin vortices and an antiferromagnetic order parameter perpendicular to the magnetizations of the ferromagnetic layers (**b**). The arrows indicate the direction of the corresponding order parameter. The rotation of the antiferromagnetic order parameter occurs in the plane of the layers

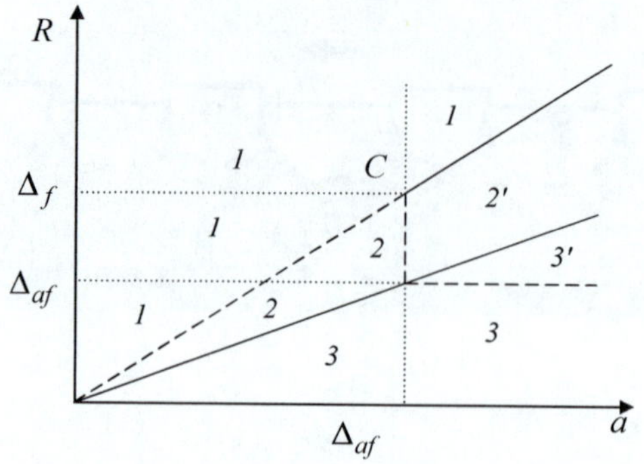

Fig. 7.10 Phase diagram of the spin-valve system: 1—phase with 90° domains in the ferromagnetic layers, 2—phase with single-domain ferromagnetic layers and the exchange spirals in antiferromagnetic interlayer (Slonczewski phase), 2′—phase with 90° domains in the antiferromagnetic layer, 3—phase with static 90^0 spin vortices and antiferromagnetic order parameter perpendicular to magnetizations of ferromagnetic layers, 3′—phase with 90° exchange spirals, parallel to interfaces in the antiferromagnetic layer. The solid lines correspond to the lines of first order phase transitions, with C being the critical point

phase corresponds to phase 1 with 90° domains. Phase 7 is missing, and the area of its existence is occupied by phase 5 (in the new numbering—3).

In phase 3′, instead of 180° exchange spirals in areas of the same type, as already mentioned in Sect. 4.4, 90° exchange spirals are formed in both styles, parallel to the interface and differing in the direction of rotation in areas of different types. Away from the boundary, the antiferromagnetic order parameter is orthogonal to the magnetizations of the ferromagnetic layers.

In the case when the surface power of two 90° domain walls in an antiferromagnet is lower than one 180° domain wall, instead of phase 4, phase 2′ is realized, in which the magnetizations of the ferromagnetic layers are mutually orthogonal, and the antiferromagnetic layer is divided into domains by the domain walls. Their number is twice as large as in the case shown in Fig. 7.7 (Fig. 7.11). The relations (7.14) and (7.15), which determine the boundaries between the phases, practically do not change. The transition from phase 2 to phase 2′ occurs smoothly as the layer thickness increases. The line of phase transitions between phases 1 and 2′ ends at the critical point C at $a \sim \Delta_{af}$ (Fig. 7.10).

7.4 Phase Transition "Polydomain State—Monodomain State with Exchange Spirals" in the Antiferromagnetic Layer

Consider the transition between phases 4 and 6 in Fig. 7.6 in more detail. The transition between the phases takes place at a certain critical value of the ratio a/R. In the

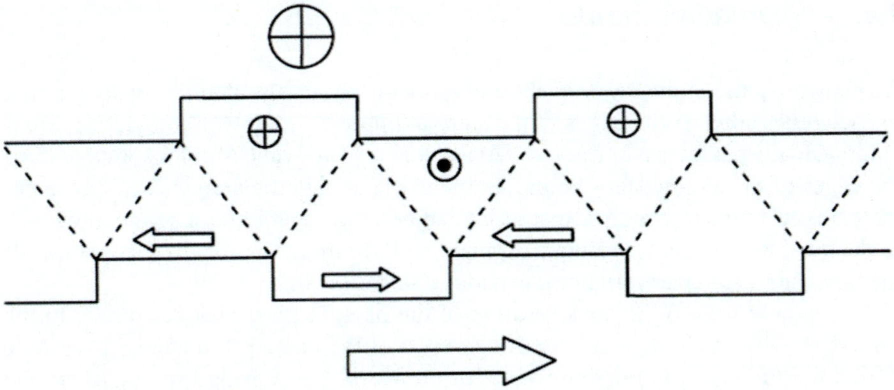

Fig. 7.11 90° domain walls in the antiferromagnetic layer. The arrows indicate the direction of the corresponding order parameter

previous section, it was estimated as $(3/4)^{1/2}$ under the assumption that the power of the exchange spiral at the interface of the layers is equal to the power of the Bloch domain wall in the antiferromagnet. That did not take into account the fact that the exchange spiral is curved towards the antiferromagnetic layer.

Indeed, because the power of the order parameter inhomogeneity is less in the antiferromagnet than in the ferromagnet, the exchange spiral tends to be located entirely in the antiferromagnetic layer. However, its edges must coincide with the edges of the neighboring atomic steps at the interface. Therefore, the displacement of the middle part of the exchange spiral into the antiferromagnet is accompanied by an increase in its surface and, consequently, the surface power. The behavior of the exchange spiral is similar to the conduct of an elastic membrane fixed at the edges and located in a potential relief.

This section is devoted to studying the deflection of the emerging exchange spirals and the refinement of the parameter $\frac{a}{R}$ corresponding to the phase transition in the antiferromagnetic layer. In a real crystal, the spread over the width of the steps R leads to a smearing of the first-order phase transition we are studying. Therefore, to identify the parameters corresponding to this transition, we considered an idealized system in which the atomic steps parallel to the axis y of the orthogonal coordinate system at each interface are located periodically with a period R, even steps leading to an increase and odd steps leading to a decrease in the thickness of the interlayer by one atomic plane (Fig. 7.7). The steps at the opposite boundaries of the interlayer are shifted relative to each other along the axis x by a distance $R/2$. Thus, a two-dimensional problem with periodic along x boundary conditions was solved.

7.4.1 Simulation Results

As shown by the simulation, in the polydomain phase, the domain walls parallel to each other connect the edges of the steps at opposite interlayer boundaries. Their length L in the plane xz is $\sqrt{R^2 + a^2}$. At a distance of an order Δ_{af} of magnitude from the edges of the atomic steps bounding the domain wall, the domain wall thickness increases from the interatomic distance to an order of magnitude Δ_{af}, and in the rest of the area, the wall is a traditional domain wall. In this case, the magnetizations of the ferromagnetic layers are homogeneous and collinear.

Let us now turn to the consideration of the phase with exchange spirals. In the exchange spiral, there is a cumulative reversal of the order parameters by an angle π. Let the function $\xi(x)$ define the set of points of the exchange spiral, where the angle $\theta_{\mathrm{af}} = \pi/2$ and the maximum value $\xi(x)$ deflect the exchange spiral d.

Since the studied phase transition is a transition of the first kind, the polydomain and monodomain phases with exchange helices coexist in a wide range of layer thicknesses. The total power of the spin-valve structure in these phases was calculated. The implementation of a particular phase was ensured by the appropriate choice of the initial spin distribution. The dependence of the power difference of the poly- and monodomain phases on the ratio $\zeta = L/R$ is shown in Fig. 7.12. It is easy to see that the phase transition corresponds to a value of $\zeta = 1.84$, and not $\zeta = 1$, as initially assumed. A change in the anisotropy constant $K_{||}$ in the range of $0.001J_f$–$0.1J_f$ led to a change in ζ by no more than 10% [5].

The dependence of the deflection of the exchange spiral on the anisotropy constant and its shape obtained as a result of modeling are shown in Figs. 7.13 and 7.14, respectively.

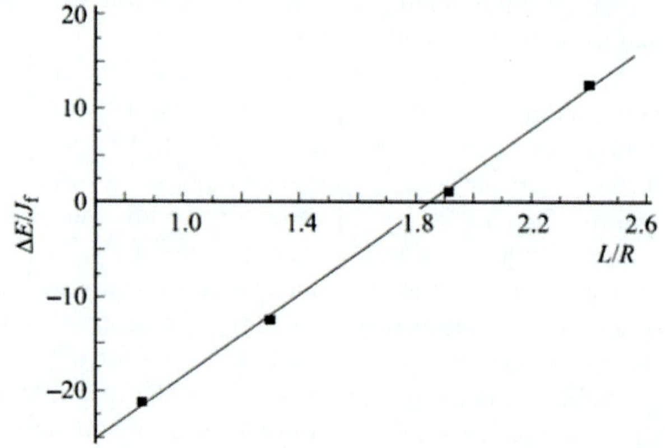

Fig. 7.12 Dependence of the power difference of the polydomain and monodomain phases on the L/R parameter

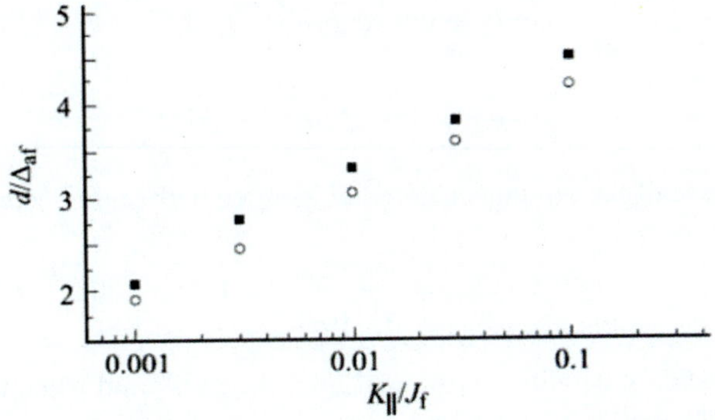

Fig. 7.13 Dependence of the deflection of the exchange spiral on the anisotropy constant: results of numerical modeling in the case $\frac{J_{af}}{J_f} = -0.1$, $\frac{J_{f,af}}{J_f} = -0.1$, $R = 200$—squares; calculation in the framework of the analytical model-circles

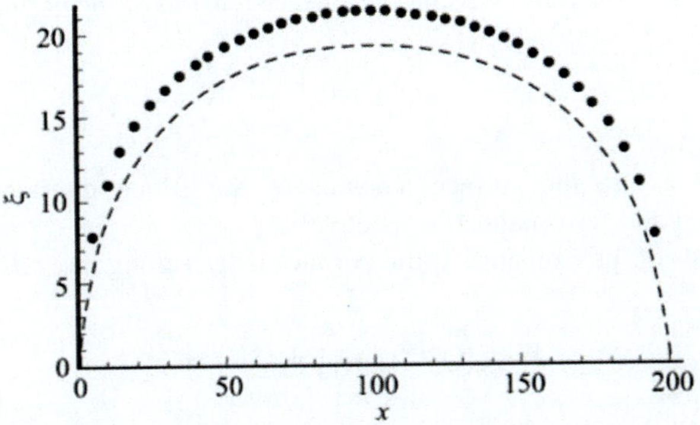

Fig. 7.14 Shape of the static exchange spiral: numerical modeling results in the case $\frac{J_{af}}{J_f} = -0, 1$ of $\frac{J_{af}}{J_f} = -0, 1$, $K_{||}/J_f = 0, 01$—points; calculation in the framework of the analytical model-a dashed line

7.4.2 *Analytical Model*

For an analytical study of the shape of the exchange spiral at the interface of the layers, we consider the following simplified model. Due to the inequality $d \ll R$, we can assume that the surface tension of the exchange spiral at a given point is a function of the local value $\xi(x)$. To find this function, we consider the exchange helix in the continuum approximation, as was done in Chap. 5 for the case of the exchange spiral at the compensated interface of the layers. Most of the exchange spiral is in the antiferromagnet, and the smaller part is in the ferromagnet. In the absence of an external magnetic field, the angle of rotation of the order parameters in the plane of the layers, calculated from the easy axis, is described by the following functions [5]:

$$\cos\theta_{\rm f} = {\rm th}[-\sqrt{2\alpha_{\rm f}}(z - z_0)], \tag{7.16}$$

$$\cos\theta_{\rm af} = {\rm th}[-\sqrt{2\alpha_{\rm f}}(z - \xi)], \tag{7.17}$$

where for the case of the volume centered of a lattice with eight nearest neighbors

$$\alpha_{f(af)} = \frac{K_{\|f(af)}}{4|J_{f(af)}|} \ll 1. \tag{7.18}$$

From the boundary condition $\theta_{\rm f}(0) = \theta_{\rm af}(0)$, which is valid when the anisotropy constant is much smaller $J_{\rm f,af}$, we have

$$z_0 = \xi\sqrt{\frac{\alpha_{\rm af}}{\alpha_{\rm f}}}. \tag{7.19}$$

Integrating the z software, we find the surface tension of the domain wall

$$\widetilde{\varepsilon}(\xi) = \frac{1}{2}\left[(\varepsilon_{\rm af} + \varepsilon_{\rm f}) + {\rm th}\left(\frac{\xi}{\Delta_{\rm af}}\right)(\varepsilon_{\rm af} - \varepsilon_{\rm f})\right], \tag{7.20}$$

where $\varepsilon_{\rm af}$ and $\varepsilon_{\rm f}$ are the surface tensions of the Bloch domain walls in the antiferromagneti and ferromagnet, respectively.

The power W of the exchange spiral per unit length along the axis y is equal to

$$W = \int_0^R {\rm d}x\tilde{\varepsilon}(\xi)\sqrt{1 + (\xi'(x))^2}. \tag{7.21}$$

Here we have neglected the change in the surface tension of the sections of the exchange spiral located at a distance of about an order $\Delta_{\rm af}$ of magnitude from the edges of the atomic steps. The boundary conditions have the form

$$\xi(0) = \xi(R) = 0. \tag{7.22}$$

By varying the functional (7.21) by ξ, we obtain the Euler equation

$$\frac{\partial\tilde{\varepsilon}(\xi)}{\partial\xi} - \frac{\tilde{\varepsilon}(\xi)\xi''}{1 + (\xi')^2} = 0 \tag{7.23}$$

Neglecting the summand $(\xi')^2 \ll 1$, and assuming that $\tilde{\varepsilon}(\xi) \approx \varepsilon_{\rm af}$, we arrive at the equation

$$\xi'' + \frac{\varepsilon_{\rm f} - \varepsilon_{\rm af}}{2\varepsilon_{\rm af}\Delta_{\rm af}}{\rm ch}^{-2}\left(\sqrt{2\alpha_{\rm af}}\xi\right) = 0. \tag{7.24}$$

Performing variable substitution

$$\eta = \sqrt{2\alpha_{\mathrm{af}}}\xi \equiv \xi/\Delta_{\mathrm{af}} \tag{7.25}$$

and

$$q = \frac{x}{\Delta_{\mathrm{af}}}\sqrt{\frac{\varepsilon_{\mathrm{f}} - \varepsilon_{\mathrm{af}}}{\varepsilon_{\mathrm{af}}}}, \tag{7.26}$$

we come to the equation [2]

$$2\eta_{qq}^{''} = -\frac{1}{\mathrm{ch}^2\eta}, \tag{7.27}$$

the first quadrature of which has the form

$$(\eta_q^{'})^2 = -\mathrm{th}\eta + A \tag{7.28}$$

At the point $x = R/2\eta_q^{'} = 0$, and the value ξ reaches the maximum value d. Then $A = \mathrm{th}\left(\frac{d}{\Delta_{\mathrm{af}}}\right) = \mathrm{th}\eta_{max}$ and

$$\eta_q^{'} = \pm\sqrt{\mathrm{th}\eta_{max} - \mathrm{th}\eta}. \tag{7.29}$$

On the plot $0 < x < \frac{R}{2}$, the value $\eta' > 0$. Since $\eta(0) = 0$ (7.22), then

$$q = \int\limits_0^{\eta} \frac{\mathrm{d}p}{\sqrt{\mathrm{th}\eta_{\max} - \mathrm{th}p}}, \tag{7.30}$$

and the deflection of the exchange spiral is found from the equation

$$\int\limits_0^{\eta_{\max}} \frac{\mathrm{d}p}{\sqrt{\mathrm{th}\eta_{\max} - \mathrm{th}p}} = q_{\max}, \tag{7.31}$$

where

$$q_{\max} = \frac{R}{2\Delta_{\mathrm{af}}}\sqrt{\frac{\varepsilon_{\mathrm{f}} - \varepsilon_{\mathrm{af}}}{\varepsilon_{\mathrm{af}}}}. \tag{7.32}$$

The dependence of the deflection of the exchange spiral on the anisotropy constant and its shape obtained because of modeling are shown in Figs. 7.13 and 7.14, respectively. With an accuracy of 10%, they coincide with the results of numerical modeling.

Thus, the curvature of the exchange spin spirals leads to an increase in their surface power and a shift of the first-order phase transition point from the polydomain to the monodomain phase toward higher *L*/*R* values.

7.5 Behavior of a Spin-Valve Structure in a Magnetic Field in the Region of Applicability of the Exchange Approximation

The process of demagnetization of ferromagnetic layers in phases 5 and 7 in Fig. 7.6 occurs almost independently, and the hysteresis loop must coincide with that in a two-layer system consisting of one ferromagnetic and one antiferromagnetic layer. Here and further on, we assume that the maximum magnetic field is much smaller than the exchange field in an antiferromagnet. Therefore, the magnetization of antiferromagnetic layers can be neglected.

In phase 3, in a weak magnetic field that exceeds the anisotropy field in the plane of the ferromagnetic layers, the magnetizations of the ferromagnetic layers are oriented at an angle of 45° to the field, remaining almost perpendicular to each other. In this case, the magnetization of the system is $M_{\max}/\sqrt{2}$, where $M_{\max}$ the maximum magnetization of the ferromagnetic layers is. Its further evolution can be investigated by minimizing the sum of the interaction energies of the ferromagnetic layers with each other [formulas (4.20) or (7.10)] and with the magnetic field. The power of ferromagnetic layers in an external magnetic field with induction *B* is given by the formula

$$W_{\mathrm{f}} = -2MaBb(\delta_1 + \delta_2)\cos\frac{\psi}{2} \tag{7.33}$$

For the angle ψ between the magnetizations of the layers, we obtain, in the case $\frac{\gamma_{\mathrm{f,af}}a}{\gamma_{\mathrm{af}}} \gg 1$ at $\sigma_1 = \sigma_2$, the transcendental equation

$$\frac{J_{\mathrm{af}}}{ab^2}\left(\frac{\pi}{2} - \psi\right) = MaBb\sin\frac{\psi}{2} \tag{7.34}$$

The characteristic field B^*, at which there is a significant change in the magnetization, is

$$B^* \sim \frac{J_{\mathrm{af}}}{Ma^2b^3} \tag{7.35}$$

Not too close to T_N, its value is much smaller than the exchange field of the antiferromagnet.

If $\frac{\gamma_{\mathrm{f,af}}a}{\gamma_{\mathrm{af}}} \ll 1$, then

$$tg\frac{\psi}{2} = \frac{J_{f,af}}{J_{f,af} + 2Mab^3 B} \quad (7.36)$$

and the characteristic field B^* is equal to

$$B^* \sim \frac{J_{f,af}}{Mab^3} \quad (7.37)$$

In phase *1*, in a weak magnetic field, the domains of the first type with a parallel orientation of the magnetizations of the layers are oriented along the field. In this case, the magnetization of the system is $M_{\max}/2$. The magnetizations of ferromagnetic layers in domains of the second type with an antiparallel orientation of the magnetizations in the zero-field behave similarly to the magnetizations of sublattices in a bulk antiferromagnet. They are oriented almost perpendicular to the external field.

As the magnitude B increases, the angle Ψ between them decreases. The characteristic value of induction $B*$ of the external magnetic field, in which the angle Ψ changes significantly, is determined similarly to phase 3 and is given in order of magnitude by the formulas (7.35) and (7.37). Therefore, the shape of the hysteresis loops in phases 1 and 3 (Fig. 3.4b) differs only in the magnitude of the magnetization in weak fields.

7.6 Comparison with the Experiment. Magnetic Phase Diagram of a Spin-Valve Structure with an Antiferromagnetic Oxide Layer

A number of papers have appeared, see, e. g., the review [M. Finazzi, et al. *Surf. Sci. Reports* **64**, p.139–167 (2009)], which assess magnetic nanostructures with antiferromagnetic oxide layers CoO and NiO on a massive substrate Fe(001). Because these layers are dielectric, in contrast to the ferromagnetic metal layers, these structures can be used to create devices based on the tunnel magnetoresistance effect [3].

In some works [A. Brambilla et al. *Phys. Rev. B* **72**, 174,402 (2005); N. Rougemaille et al. *Phys. Rev. B* **76**, 214,425 (2007); A. Brambilla et al. *Phys. Rev. B* **79**, 172,401 (2009)], the structure of one or two wedge-shaped layers (antiferromagnetic and/or ferromagnetic) deposited on a thick ferromagnetic substrate is investigated. Moreover, the thicknesses of the two wedge-shaped layers varied in two mutually perpendicular directions in the plane of the layer (Fig. 7.15).

The phase diagram of the structure shown in Fig. 7.15 differs significantly from the phase diagram of a spin-valve system with layers of the same thickness. Therefore, the purpose of this section is a theoretical consideration of the magnetic phase diagram of this structure and a comparison of the results obtained with experimental data. We will assume that the atomic planes of the antiferromagnetic oxide parallel to the substrate are uncompensated. There are reasons for this, since, according to the latest experimental data, the nickel oxide on the Fe(001) substrate has the orientation of

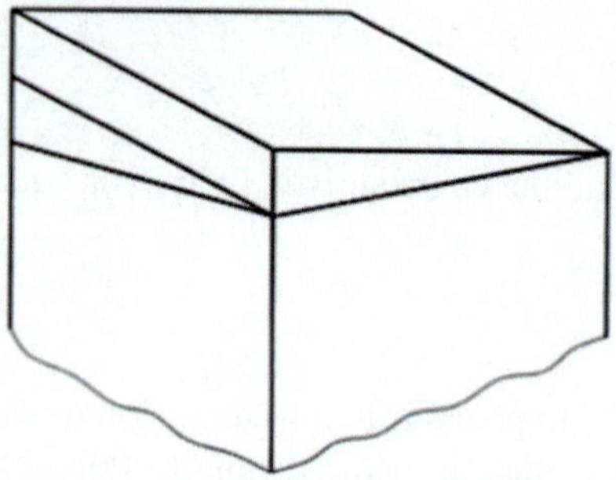

Fig. 7.15 Three-layer wedge-shaped structure

the layers (111). Since the antiferromagnetic ordering of *NiO* and CoO is described by the wave vector $\left(\frac{1}{2}, \frac{1}{2}, \frac{1}{2}\right)$, the planes (111) are uncompensated.

In the future, we will limit ourselves to the applicability of the exchange approximation when the characteristic width of the atomic steps R at the interface and the thickness of two thin wedge-shaped layers are much smaller than the thickness of the traditional domain walls in ferromagnets and antiferromagnets. In this case, the contribution of the power of the single-ion anisotropy is insignificant.

In the case of a thick ferromagnetic layer in the three-layer structure, it can be argued that it always remains in the monodomain state. Thus, along with the parameter R, there are two other geometric parameters: the thickness of the antiferromagnetic layer d_{af} and the thin upper ferromagnetic layer thickness.

7.6.1 *Phase Diagram*

We begin our consideration of the magnetic phase diagram with the case when the parameter R far exceeds all other characteristic parameters of the problem, particularly the thickness of the layers d_{af} and d_f (while remaining within the exchange approximation), and the thin ferromagnetic layer is in a polydomain state. In each domain, the optimal orientation of the order parameters is implemented.

In a spin-valve structure with the same layer thicknesses, only one type of domain wall occurs in the area of the validity of the exchange approximation at $|J_{af}| \sim J_{f,\,af}$ (Fig. 7.2). In the case under consideration, the number of possible types of domain walls is significantly higher. Moreover, it is crucial at which interface the atomic step is located. If it is located on the border with a thick ferromagnetic layer, the 180° domain wall penetrates two layers: an antiferromagnetic layer and a thin ferromagnetic layer. If the step is located on the border with a thin ferromagnetic layer, then 180° the domain wall intersects only this layer. In the interlayer, distortions of the antiferromagnetic order parameter occur in the area of the existence of such a domain wall. However, outside this area, on different sides of the atomic step, the directions of the antiferromagnetism vector coincide.

The structure of the domain wall depends significantly on the thickness of the layers. Let us consider as a variable d_{af} the thickness of the antiferromagnetic layer. As will be shown below, there are three value ranges of d_{af} value in which the domain walls generated by frustration have different features.

Let's first consider the smallest values d_{af}. The domain wall, originating in the interlayer at the edge of the atomic stage, has a width at the interface δ_0. In this area, near the edge of the atomic step, the mutual orientation of the magnetic order parametersdiffers from the optimal one (parallel on one side of the step and antiparallel on the other). Due to this, the power of the interlayer interaction per unit length of the step increases by

$$W_{f,\,af} \sim \frac{J_f}{b}\gamma_{f,af}\delta_0. \tag{7.38}$$

The nascent domain wall expands as it moves away from the interface. The thickness of the domain wall grows to an order of magnitude d_{af}. After that, the lines of constant value θ_{af} run parallel to the boundaries of the layers, occupying the area $d_{af} < |x| < \delta_{0f}$, where δ_{0f} is the thickness of the domain wall in a thin ferromagnetic layer at the ferromagnetic-antiferromagnetic interface, and the atomic step is located at the point $x = 0$. The domain wall thickness continues to grow in a thin ferromagnetic layer, reaching an order of magnitude d_f at its free boundary.

The values δ_0 and δ_{0f} can be found from simple power considerations. The additional power associated with the inhomogeneity of the order parameter in the antiferromagnetic layer is calculated as the value per unit length of the edge of the atomic step

$$W_{af} \sim \frac{J_f}{b}\gamma_{af}\left[\ln\frac{\mathrm{d}_{af}}{\delta_o} + \frac{\delta_{of} - \mathrm{d}_{af}}{\mathrm{d}_{af}}\right]. \tag{7.39}$$

The similar power of the magnetization distortion in a thin ferromagnetic layer is equal to

$$W_f \sim \frac{J_f}{b}\ln\frac{\mathrm{d}_f}{\delta_{0f}}. \tag{7.40}$$

Minimizing the total power $W_{f,\,af} + W_{af} + W_f$ by the parameters δ_0 and δ_{0f}, we obtain:

$$\delta_o \sim \gamma_{af}/\gamma_{f,af}, \tag{7.41}$$

$$\delta_{0f} \sim d_{af}/\gamma_{af}. \tag{7.42}$$

The condition $\delta_{0f} < d_f$ gives a limit on the thickness of the antiferromagnetic layer

$$d_{af} < \gamma_{af} d_f. \tag{7.43}$$

In the area of the values of the thicknesses of the antiferromagnetic layer $d_{af} > \gamma_{af} d_f$, the magnitude $\delta_{0f} > d_f$ and broadening of the domain wall in a thin ferromagnetic layer can be neglected. In this case, the power w_f has the form

$$w_f \sim \frac{J_f d_f}{b\delta_{0f}} \tag{7.44}$$

what gives the value

$$\delta_{0f} = (d_f d_{af}/\gamma_{af})^{\frac{1}{2}}. \tag{7.45}$$

The range of values d_{af} in which these formulas are valid is bounded from above by an inequality $d_{af} < \delta_{0f}$ leading to the condition

$$d_{af} < d_f/\gamma_{af}. \tag{7.46}$$

At large thicknesses d_{af}, it becomes essential at which interface the atomic step is located. The structures of the domain walls generated by the steps at different interfaces differ significantly.

If the step is located on the boundary between the thin ferromagnetic layer and the interlayer, then the distortions of the antiferromagnetic order parameter no longer capture the entire thickness of the interlayer, and the domain wall takes the form of a domain wall in the two-layer system discussed in the fourth chapter (Fig. 4.3). Its thickness in the ferromagnetic layer $\delta_{0f}^{(1)}$ is equal to [5]

$$\delta_{0f}^{(1)} \sim d_f/\gamma_{af}. \tag{7.47}$$

If the domain wall is generated by an atomic step at the boundary between a thick ferromagnetic layer and an interlayer, then within the exchange approximation

$$\delta_{0f}^{(2)} \sim d_{af}. \tag{7.48}$$

Distortions of the antiferromagnetic order parameter capture the entire thickness of the interlayer, and, since $\delta_{0f}^{(2)} \gg \delta_{0f}^{(1)}$, domain walls with a thickness $\delta_{0f}^{(1)}$ can exist against the background of a wall with a thickness of $\delta_{0f}^{(2)}$.

Since in the case $d_f, d_{af} \sim 1$ of nm, the values of the thicknesses of the domain walls are also several nanometers, the nanodomain state is realized in a thin ferromagnetic layer at a distance between the edges of the atomic steps of the order of 10 nm, which is much greater than the thickness of the domain walls. The region of its existence is divided into three subregions: $d_{af} < \gamma_{af} d_f$; $\gamma_{af} d_f < d_{af} < \frac{d_f}{\gamma_{af}}$; $d_{af} > d_f/\gamma_{af}$ (phases 1, 2, and 3 in Fig. 7.16, respectively), in which domain walls of

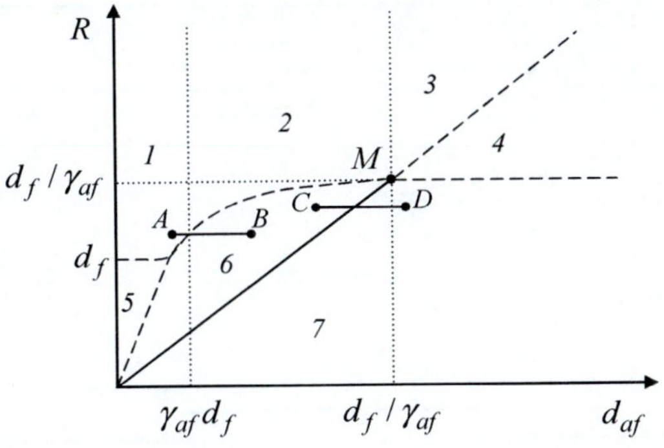

Fig. 7.16 Phase diagram of a three-layer system: 1-4-nanodomain phases, in which domain walls of different types are observed; 5-phase with static spin vortices in a thin ferromagnetic layer; 6-phase with exchange spirals in an antiferromagnetic layer; 7—collinear phase; A solid line corresponds to a first order phase transition, *M*— critical point. The segments AB and CD show the interval of the experimental data of the works

different types are observed. The transition between them, that is, transforming the domain walls during $d_{\rm af}$ change occurs smoothly and is not a phase transition.

We now find the boundary of the existence of the nanodomain state by reducing the characteristic distance between the edges of the steps R. In the subregion $d_{\rm af} < \gamma_{\rm af} d_{\rm f}$, the domain walls begin to overlap at $R \sim d_{\rm f}$. In the range of values $\delta_{of} \ll R \ll d_{\rm f}$ in the thin ferromagnetic layer near the boundary with the antiferromagnetic layer, static spin vortices similar to those described in Chap. 4 will arise. In the remaining part of the layer, its magnetization will turn perpendicular to the magnetization of the thick ferromagnetic layer to reduce the power of the spin vortices (phase 5 in Fig. 7.16).

In the case $d_{\rm af} \ll R \ll \delta_{0\rm f}$, the magnetization of the thin ferromagnetic layer becomes almost uniform, and 90° exchange spirals are formed in the antiferromagnetic layer. That is phase 6 in Fig. 7.16—the Slonczewski phase.

In the case $R \ll d_{\rm af}$, static spin vortices are formed in the interlayer near each of the layer interfaces (Fig. 7.9). In the rest of the interlayer volume, the antiferromagnetic order parameter is oriented perpendicular to the magnetizations of the ferromagnetic layers, which become collinear (phase 7 in Fig. 7.15). And if the previous transitions from phase 1 to phase 5 and phase 5 to phase 6 with a decrease R occurred continuously and were not, strictly speaking, phase transitions, the transition from phase 6 to phase 7 is a real first order phase transition, in which the orientation of the magnetizations of ferromagnetic layers abruptly changes from mutually perpendicular to the collinear.

What happens when the parameter R decreases in the subregion $\gamma_{\rm af} d_{\rm f} < d_{\rm af} < d_{\rm f}/\gamma_{\rm af}$? Since in this subdomain $\delta_{0\rm f} \gg d_{\rm f}$, the overlapping of the domain walls at $R \sim \delta_{0\rm f}$ leads to a smooth transition from phase 2 immediately to the Slonczewski phase 6. At $R \sim d_{\rm af}$ the same time, a phase transition occurs from phase 6 to phase 7.

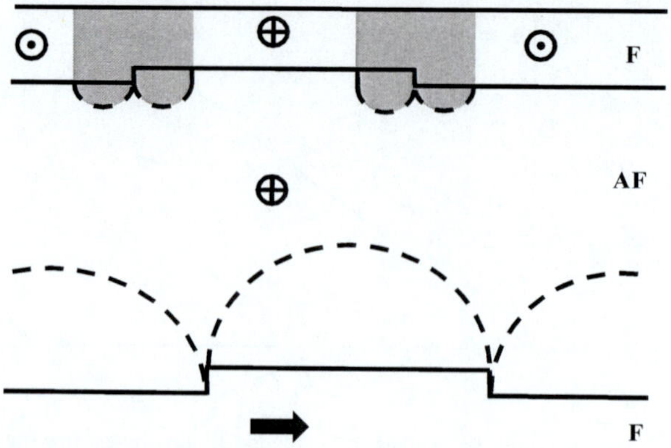

Fig. 7.17 Orientation of the order parameters in phase 4

In the area $d_{af} > d_f/\gamma_{af}$ with a decrease R, the thick walls $\delta_{0f}^{(2)}$ generated by the steps at the boundary with the thick ferromagnetic layer will first begin to overlap. When $R < d_{af}$, static spin vortices are formed in the interlayer near the interface with a thick ferromagnetic layer and in the remaining volume of the interlayer, the antiferromagnetism vector is oriented perpendicular to the magnetization of the thick ferromagnetic layer (Fig. 7.17). The thin ferromagnetic layer is in the polydomain phase due to the domain walls generated by the atomic steps at its boundary with the interlayer (phase 4 in Fig. 7.16). The magnetization orientation is parallel in domains of one type, and in domains of another type it is antiparallel to the antiferromagnetism vector.

Since the transition from phase 3 to phase 4 R occurs smoothly as it decreases, there is a critical point M in the phase diagram, at which the line of phase transitions from phase 6 to phase 7 ends.

With a further decrease R in $R \sim \delta_{0f}^{(1)}t$, the domain walls overlap in a thin ferromagnetic layer, and a smooth transition to phase 7 occurs. In variables (R, d_f), the phase diagram we described is shown in Fig. 7.18.

7.6.2 Comparison with the Experiment

The behavior of multilayer structures observed in the experiment is qualitatively consistent with the given phase diagrams. The areas of variation of the parameters d_f and d_{af} corresponding, in our opinion, to the experimental data of the above works are shown in Figs. 7.16 and 7.18.

The nanodomain state was observed in three-layer systems: Fe/NiO/Fe [N. Rougemaille et al. *Phys. Rev. B* **76**, 214,425 (2007)] and Fe/CoO/Fe [A. Brambilla et al. *Phys. Rev. B* **79**, 172,401 (2009)].

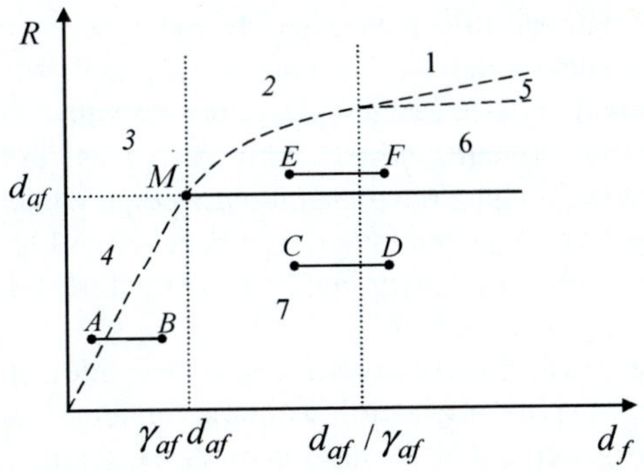

Fig. 7.18 Phase diagram of a three-layer system: 1-4-nanodomain phases, in which domain walls of different types are observed; 5-phase with static spin vortices in a thin ferromagnetic layer; 6-phase with exchange spirals in an antiferromagnetic layer; 7—collinear phase; a solid line corresponds to a first order phase transition, *M*—critical point

In the Fe/NiO/Fe(001) system, considered in [A. Brambilla et al. *Phys. Rev. B* **72**, 174,402 (2005)], with an increase in the thickness of the interlayer at a constant value of $d_f = 1.4$ nm, a transition was observed from the mutually perpendicular orientation of the magnetizations of the ferromagnetic layers to their collinear orientation at $d_{af} = 4$ nm, which corresponds to the transition from phase 6 to phase 7 (segment CD in Fig. 7.16). Since at the transition point $R \sim d_{af}$, we get an estimate for the distance between the edges of the atomic steps $R \sim 4$ nm.

In the same work, the parameter d_f was changed at constant values d_{af}. At $d_{af} <$ 4 nm, the sample was in phase 6, and a change d_f from 1.4 to 5.5 nm did not lead to a transition to another phase (EF segment in Fig. 7.18). At $d_{af} > 4$ nm, the sample was in phase 7, and a change of d_f from 1.4 to 4.5 nm did not lead to a transition to another phase (segment CD in Fig. 7.18), which fully corresponds to the proposed phase diagram.

In the work [N. Rougemaille et al. *Phys. Rev. B* **76**, 214,425 (2007)], the Fe/NiO/Fe(100) system was studied, and the thickness of the thin ferromagnetic layer d_f varied in the range of 1.3–6 nm with a constant value $d_{af} = 4$ nm. The area of the studied values d_f began in phase 4 ($d_f = 1.3$ nm) and crossed the collinear phase 7 (segment AB in Fig. 7.18) boundary, which led to a blurring of the nanodomain partitioning pattern.

In the work [A. Brambilla et al. *Phys. Rev. B* **79**, 172,401 (2009)], three different states were observed in the multilayer structure of Fe/CoO/Fe at $d_f = 1.4$ nm in the range of interlayer thicknesses of 1 nm $< d_{af} <$ 5 nm: the nanodomain state at 1.5 nm $< d_{af} <$ 4 nm, the collinear state of the magnetizations of ferromagnetic layers at smaller interlayer thicknesses, and the mutually perpendicular orientation of the magnetizations at thicknesses $d_{af} > 4$ nm. The CoO layer differs from the NiO layer by a smaller parameter value γ_{af}.

The only fact that is not interpreted in the framework of our theory is the absence of domains in the Fe/CoO/Fe system in the range of small values of the interlayer thicknesses can be explained by the presence of ferromagnetic punctures in the interlayer (pinholes), which our model does not take into account. With a more significant layer thickness, a nanodomain state (phase 2) occurs with the division of a thin ferromagnetic layer into 180° domains, which, with an increase in the layer thickness, passes into the Slonczewski phase 6 with a mutually perpendicular orientation of the magnetizations of the ferromagnetic layers (segment AB in Fig. 7.16). The investigated range of values d_{af} in a multilayer Fe/CoO/Fe structure did not reach the phase transition line to the collinear phase 7.

Let us focus on the possibilities of experimental verification of the predicted phase diagram. The method of spin-polarization scanning tunneling microscopy allows us to study domain walls generated by frustration on the structure's surface. Therefore, it is possible to study phases 1–4, in which the upper thin ferromagnetic layer is divided into domains. At the same time, the detection of two types of domain walls with significantly different thicknesses in phase 3 and the transition from phase 3 to phase 4 with one type of domain walls and domain magnetizations deployed by 90° concerning phase 3 is of the most significant interest.

Unfortunately, we do not know direct methods for studying the distributions of order parameters near the interface of layers and, in particular, static spin vortices. The mutual orientation of the magnetizations of ferromagnetic layers can be studied using the magnetooptical Kerr effect [1].

References

1. Morosov AI, Sigov AS (1999) Phase diagram of multilayer structures ferromagnet-layered antiferromagnet. FTT 41(7):1240–1247
2. Levchenko VD, Morosov AI, Sigov AS (2002) Phase diagram of multilayer magnetic structures. ZhETF 121(5):1149–1156
3. Morosov AI, Morosov IA (2008) Magnetic phase diagram of the spin-valve structure of ferromagnet-antiferromagnet-ferromagnet with rough interfaces. FTT 50(10):1846–1852
4. Morosov AI, Morosov IA (2011) Magnetic phase diagram of a spin-valve structure with an antiferromagnetic oxide layer. FTT 53(1):58–65
5. Morosov AI, Markets DO (2012) The phase transition "polydomain state—monodomain state with exchange helices" in the antiferromagnetic layer of the spin-valve structure. FTT 54(4):715–719

Chapter 8
Surface Spin-flop Transition in Antiferromagnet

8.1 Volume Spin-flop Transition

It is well known that the physical properties near the surface differ from those in its bulk. This difference can lead to the fact that the phase transition near the crystal surface will occur at a value of temperature, external field, or other control parameters that do not coincide with the values corresponding to the phase transition in the crystal volume.

One example of such a difference is the overturning of sublattices of a uniaxial collinear two-sublattice antiferromagnet in a magnetic field applied parallel to the easy axis, the so-called spin-flop transition. In the literature, as a rule, theoretical models of surface transitions consider a smooth flat surface of a crystal. The actual surface of a crystal always contains atomic steps that change the coordinate corresponding to the position of the flat surface by the height of one atomic layer. In the case of an uncompensated surface of an antiferromagnet, atomic steps lead to the partition of the surface into two types of areas. The magnetization vector of the upper layer of atoms is parallel and antiparallel to the external magnetic field, respectively. The spin-flop transition in different types of areas occurs according to different scenarios, which also depend on the relationship between the correlation radius of the antiferromagnetic order parameter and the size of these areas. Exploring these scenarios is the first goal of this chapter.

The second goal is to consider spin-flop, and spin-flip transitions in plane-parallel layers of an antiferromagnet in the nanometer thickness range under conditions when the correlation radius of the order parameter is comparable to the layer thickness and the size effects become significant. The model of a plane-parallel layer of an antiferromagnet allows us to satisfactorily describe antiferromagnetic atomic chains on the surface of a ferromagnet and multilayer magnetic structures with antiferromagnetic interaction between neighboring ferromagnetic layers.

For the first time, the spin-flop transition in the volume of a boundless antiferromagnet was predicted in the pioneering work by Neel [Neel. *Ann. Phys.* **5**, 232 (1936)]. Consider a system of quasi-classical localized spins at a temperature

A. S. Sigov, *Multilayer Magnetic Nanostructures*, Springer Aerospace Technology,
https://doi.org/10.1007/978-981-19-6246-2_8

$T \ll T_N$ (where T_N is the Neel temperature), when the modules of the localized spins can be assumed to be unchanged. The power of the exchange interaction in the nearest neighbor interaction approximation has the form [1]

$$W_{ex} = -\frac{1}{2}\tilde{z} J N \cos\psi, \tag{8.1}$$

where $\tilde{z}$ is the number of nearest neighbors, $J < 0$ is the exchange integral, the value of the spins is included in the corresponding constants (in this case in J), N is the number of spins, and ψ is the angle between the magnetizations of the sublattices.

The power of a single-ion anisotropy is equal to

$$W_{an} = -\frac{N}{2}\sum_{i=1}^{2} K \cos^2\varphi_i, \tag{8.2}$$

where $K > 0$ is the anisotropy constant, and φ_i is the angle between the magnetization *of the* i-th sublattice ($i = 1, 2$) and the light x-*axis* (Fig. 8.1 $\psi = \varphi_1 - \varphi_2$).

The spin power in an external magnetic field parallel to the easy axis is represented as

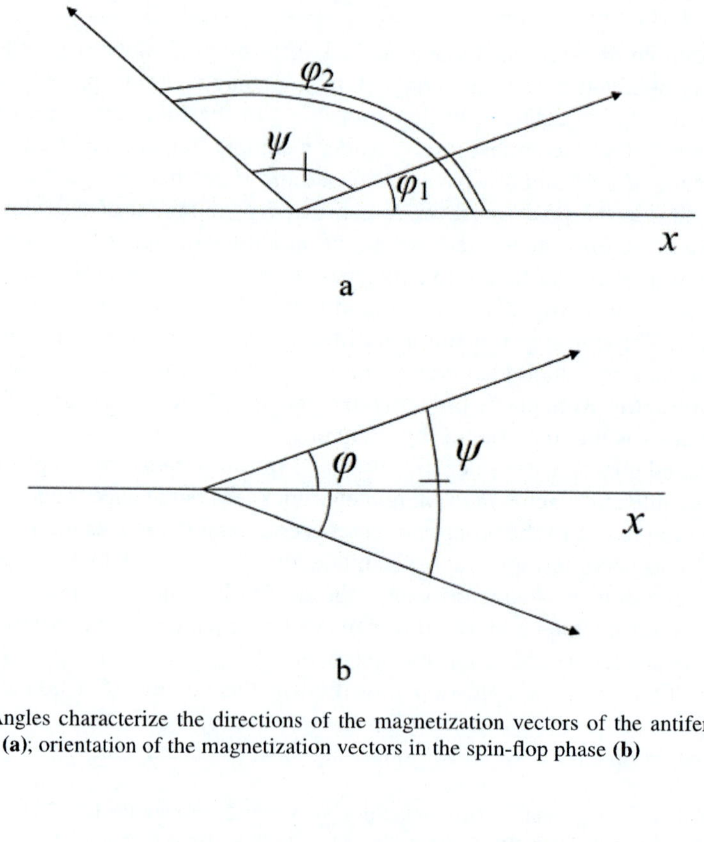

Fig. 8.1 Angles characterize the directions of the magnetization vectors of the antiferromagnet sublattices (**a**); orientation of the magnetization vectors in the spin-flop phase (**b**)

$$W_B = -\mu B_0 \frac{N}{2} \sum_{i=1}^{2} \cos\varphi_i, \tag{8.3}$$

where μ is the magnetic moment of the atom, and B_0 is the induction of an external magnetic field.

In the collinear phase, when $\varphi_1 = 0, \varphi_2 = \pi$ the total power $W = W_{ex} + W_{an} + W_B$ is equal to

$$W_1 = -\left(\frac{1}{2}\tilde{z}|J| + K\right)N. \tag{8.4}$$

In the spin-flop phase (Fig. 8.1b) $\varphi_1 = -\varphi_2 = \varphi$, and the power W is equal to

$$W_2 = -\left(\frac{1}{2}\tilde{z}J\cos 2\varphi + K\cos^2\varphi + \mu B_0 \cos\varphi\right)N. \tag{8.5}$$

Minimizing W_2 by variable φ and introducing notation

$$\alpha = \frac{K}{\tilde{z}|J|} \ll 1, \tag{8.6}$$

$$\beta = \frac{\mu B_0}{\tilde{z}|J|}, \tag{8.7}$$

finding it

$$\cos\varphi = \frac{\beta}{2(1-\alpha)}, \tag{8.8}$$

$$W_2 = -\frac{\tilde{z}|J|N}{2}\left[1 + \frac{\beta^2}{2(1-\alpha)}\right]. \tag{8.9}$$

Comparing W_1 and W_2, we obtain the magnetic field induction value corresponding to the volume spin-flop first order transition from the collinear phase to the spin-flop phase:

$$\beta_2 = 2\sqrt{\alpha(1-\alpha)} \approx 2\sqrt{\alpha}. \tag{8.10}$$

The angle φ at $\beta > \beta_1$ decreases with the growth of the field, and in the field

$$\beta_2 = 2(1-\alpha) \tag{8.11}$$

the sublattices collapse: the angle φ turns to zero. This second-order phase transition is called a spin-flip transition. When $\beta > \beta_2$, the magnetizations of the sublattices of the antiferromagnet are parallel, and we will call this phase ferromagnetic.

8.2 The Case of an Atomically Smooth Surface

Let's turn to the consideration of the surface spin-flop transition. Let us first study the atomically smooth surface.

In the case of a compensated slice, the surface phase transition does not take place, and the spins of the surface layer of atoms are tipped over in the field of the bulk spin-flop transition β_1. This is due to the atomic scale of the change in the antiferromagnetic order parameter equal to the difference in the magnetization vectors of the sublattices and the power disadvantage of such distortions. Therefore, in the future, we will only consider the uncompensated slice. The surface spin-flop transition for this case was first investigated in [D. L. Mills. *Phys. Rev. Lett.* **20,** 18 (1968)], where it is shown that the softening of the law of dispersion of surface magnons takes place in the field

$$\beta_s = \frac{\beta_1}{\sqrt{2}}. \tag{8.12}$$

In subsequent works [D. L. Mills, W. M. Saslow. *Phys. Rev.* **171,** 488 (1968); F. Keffer, H. Chow. *Phys. Rev. Lett.* **31,** 1061 (1973)], the distribution of the antiferromagnetic order parameter in the field range was found $\beta_s < \beta < \beta_1$. The authors assumed that the spin-flop orientation is realized on the surface of the antiferromagnet, which takes place in the volume of the antiferromagnet at $\beta > \beta_1$, that is, the vector of antiferromagnetism in the surface layer of atoms is perpendicular to the easy axis.

Since the volume of the antiferromagnet at $\beta < \beta_1$ is in the collinear phase, a 90° exchange spiral should appear near the surface of the antiferromagnet, the structure of which coincides with the structure of the Bloch domain wall.

The assumption that a spin-flop orientation on the surface of the antiferromagnet is superfluous since the spin distribution can be found without additional assumptions. Moreover, it is erroneous, since there is not a 90°, but a 180° exchange spiral near the surface.

After these studies, the question of the surface spin-flop transition was out of the attention of researchers for many years. A new interest in it arose after the discovery of the phenomenon of giant magnetoresistance in multilayer magnetic structures, when it became clear that the behavior in the magnetic field of multilayer structures with alternating ferromagnetic and non-magnetic layers of nanometer thickness and the antiferromagnetic exchange sign between neighboring ferromagnetic layers is largely similar to the behavior of a plane-parallel layer of an antiferromagnet with uncompensated surfaces.

Since the number of nearest neighbors for the spins located on the surface of the antiferromagnet is less than in the volume, they are more susceptible to the influence of an external magnetic field. That is why the field of the surface spin-flop transition is lower than its volume value.

Consider a slice (100) of a tetragonal volume-centered lattice. The easy axis lies in the plane of the surface. We number the uncompensated atomic planes parallel to the surface with an index j starting from the surface. Even and odd values j correspond to different sublattices. The position of the spins in the atomic plane will be set by the angle θ_j between the direction of the easy axis and the corresponding spin magnetic moment (Fig. 8.1a), and it is assumed that the magnetic moments of the atoms do not go out of their atomic plane. Then the surface magnetic moment resulting from the spin-flop transition does not have a component perpendicular to the surface. Otherwise, there is a magnetic field, which leads to an increase in the system's power.

Taking into account, the assumption made, the power of the exchange interaction takes the form

$$W_{\text{ex}} = \frac{n|J|\tilde{z}}{4} \sum_{j=1}^{\infty} \left[\cos(\theta_j - \theta_{j-1})(1 - \delta_{1,j}) + \cos(\theta_j - \theta_{j+1}) \right], \tag{8.13}$$

where n is the number of spins in the atomic plane and $\delta_{i,j}$ is the Kronecker symbol. The number of neighbors closest to a given spin that lies in the preceding or subsequent atomic plane is equal $\tilde{z}/2$. For a slice (100) of a tetragonal volume-centered lattice (with an easy axis c lying in the slice plane) $\tilde{z} = 8$.

For the single-ion anisotropy power and the Zeeman power, we have [2]

$$W_{\text{an}} = -Kn \sum_{j=1}^{\infty} \cos^2\theta_j, \tag{8.14}$$

$$W_B = -\mu B_0 n \sum_{j=1}^{\infty} \cos\theta_j, \tag{8.15}$$

the magnetic field is applied parallel to the easy axis.

Minimizing the total power W by the parameters θ_j, we obtain a system of equations

$$\begin{aligned} &\sin(\theta_j - \theta_{j-1})(1 - \delta_{1,j}) + sin(\theta_j - \theta_{j+1}) \\ &= 2\alpha \sin 2\theta_j + 2\beta \sin\theta_j. \end{aligned} \tag{8.16}$$

Numerical methods obtained the solution of this system.

It was found that if the magnetic moment of the top atomic plane in a weak magnetic field is antiparallel to the field ($\theta_1 = \pi$), then in the field β_s a surface spin-flop transition occurs and near the surface a 180° exchange spiral arises: the magnetization of the top atomic plane turns along the field, the magnetization of the second and third, fourth and fifth, etc. atomic planes in pairs are compensated. Since the magnetization of the first atomic plane is parallel to the magnetization of the last even atomic plane, the spin-flop transition results in a magnetic moment of the magnitude $2\mu n$ near the surface of the antiferromagnet 2. The reversal of the magnetization vector of the upper atomic plane by 180° can be observed by magnetic-force, spin-polarization scanning tunneling, and photoemission electron microscopy. The difficulty is that the measurements must be carried out in an external magnetic field.

The characteristic scale at which the magnetizations of the atomic planes turn in the exchange spiral under study can be estimated by considering the asymptotic behavior of the angles θ_j.

In the depth of the antiferromagnet, the values θ_j differ from their volume values (0 for even planes and π odd ones) by a small amount. That allows us to linearize the system (8.16) concerning these deviations

$$\chi_{2n} = \theta_{2n}, \; \chi_{2n-1} = \theta_{2n-1} - \pi.$$

The linearized equation has a solution

$$\chi_n = \kappa \chi_{n-1}, \tag{8.17}$$

where

$$\kappa = 1 - \sqrt{\beta_1^2 - \beta^2}. \tag{8.18}$$

It corresponds to the exponential decline of distortions deep into the antiferromagnet:

$$\chi_n = \chi_1 \exp\left[-\frac{d(n-1)}{r_c}\right], \tag{8.19}$$

where d is the distance between the nearest atomic planes, and the correlation radius of the order parameter r_c is

$$r_c = \frac{d}{|\ln|k||} = \frac{d}{\sqrt{\beta_1^2 - \beta^2}} \propto (\beta_1 - \beta)^{-\frac{1}{2}}. \tag{8.20}$$

For the first time, such a dependence of the correlation radius at $\beta < \beta_1$ was obtained in [F. Keffer, H. Chow. *Phys. Rev. Lett.* **31,** 1061 (1973)]. It is easy to see that r_c diverges in the root manner at $\beta \to \beta_1$; that is, the surface spin-flop transition propagates into the depth of the antiferromagnet as it approaches the point of the bulk spin-flop transition.

In the fields $\beta > \beta_1$, when the bulk spin-flop transition has already occurred, it is energetically advantageous for the magnetization of the upper (first) atomic plane to be directed along the field, so the 180° spiral turns into 90°. In the volume, the antiferromagnetism vector is perpendicular to the easy axis, and near the surface it is collinear to it. The overturning of sublattices in the near-surface layer occurs in more strong fields ($\beta \sim 0.25$).

It seems that in the model case of an atomically smooth surface of an antiferromagnet with $\theta_1 = \pi$ and 180° exchange spiral (the domain wal) it is energetically advantageous for the spiral to disappear, passing through the entire antiferromagnet (if this is not prevented by pinning the domain wall). Whether such a scenario is realized depends on the conditions at the opposite boundary of the sample. The plane-parallel layer of the antiferromagnet will be considered in the next section. It turns out that if the magnetization of the last atomic plane of the antiferromagnet is also antiparallel to the applied external magnetic field, then in the field of the surface spin-flop transition, all the spins of the antiferromagnet will be reversed by 180° (we are talking about the equilibrium state). When the magnetization of the last atomic plane of the antiferromagnet is parallel to the applied external magnetic field, the domain wall will be established in the center of the plane-parallel layer.

In the real case of the presence of atomic steps at the boundary of the antiferromagnet, there are areas with $\theta_1 = \pi$ and $\theta_1 = 0$ on the surface. As will be shown below, this localizes the area of inhomogeneous distortions of the antiferromagnetic order parameter near the surface of the antiferromagnet.

If the magnetic moment of the upper atomic plane in weak magnetic fields is parallel to the field ($\theta_1 = 0$), then the surface spin-flop transition is not energetically favorable, and the bulk transition occurs at $\beta = \beta_1$.

The situation in the fields $\beta > \beta_1$ is entirely similar to the case $\theta_1 = \pi$; the difference is only in the direction of rotation of the antiferromagnetic order parameter in the exchange spiral.

The characteristic scale of the exchange spiral at $\beta > \beta_1$ can be determined by the asymptotic behavior of the angles θ_j. Only now the deviations of the angles θ_j from their volume values in the spin-flop phase are small: $\chi_{2n-1} = \theta_{2n-1} - \varphi$ and $\chi_{2n} = \theta_{2n} + \varphi$, where the value φ is given by the formula (8.8).

The solution of the linearized equation has the form (8.19), where near β_1

$$\kappa = 1 - \sqrt{\beta^2 - \beta_1^2}. \tag{8.21}$$

With the growth of the field in the area $\beta > \beta_1$, the exchange spiral near the surface does not disappear, but the angle of rotation in it decreases. The dependence of the absolute value of the parameter $\chi_1 = \theta_1 - \varphi$ on the magnitude of the magnetic field is shown in Fig. 8.2.

In addition, the spiral is compressed due to a decrease in the correlation radius r_c. The parameter κ describing the asymptotic behavior of deviations χ_j in the range of fields $\beta_1 \ll \beta \ll \beta_2$ is given by the expression

$$\kappa = 1 + \frac{2\beta^2}{\beta_2^2 - 2\beta^2} \pm \sqrt{\left(1 + \frac{2\beta^2}{\beta_2^2 - 2\beta^2}\right) - 1}. \tag{8.22}$$

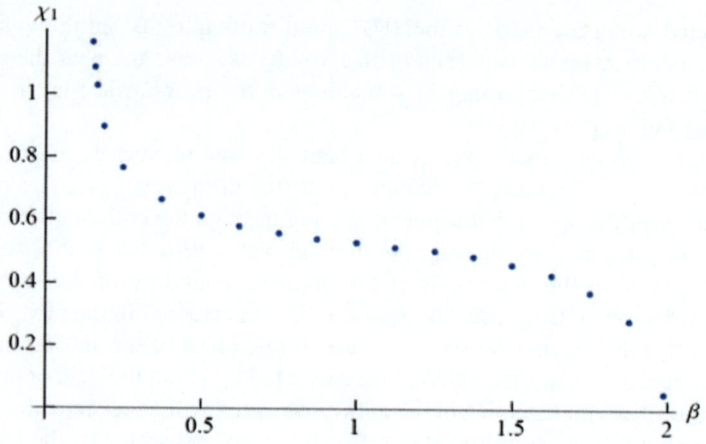

Fig. 8.2 Dependence of the angle of rotation of the magnetic moment of the surface layer of atoms on the magnetic field, larger than the field of the bulk spin-flop transition, in the case of $\alpha = 0.005$

It is easy to see that when $\beta \to \beta^* = \frac{\beta_2}{\sqrt{2}}$, the value $\kappa \to 0$, the correlation radius vanishes as well. That corresponds to the fact that at $\beta = \beta^*$ only the value θ_1 that determines the direction of magnetization of the upper atomic plane deviates from the volume value. The sign of this deviation is such that the magnetization makes with the magnetic field an angle smaller than such an angle in the volume of the crystal.

When $\beta = \beta^*$ there is a change in the sign of the value κ. If at $\beta < \beta^*$ one has $\kappa > 0$ (the minus sign in the formula (8.22)), then at $\beta > \beta^*$ κ is negative (the plus sign in the formula (8.22)).

As the magnitude of the magnetic field increases in the range $\beta^* < \beta < \beta_2$, the characteristic size of the exchange spiral begins to grow, diverging in a root type at $\beta \to \beta_2 = 0$. Since in the limit $\beta \to \beta_2$ all angles θ_j and χ_j tend to zero, the solution of the system (8.16) can be found analytically.

We introduce a dimensionless parameter

$$\Delta = 1 - \frac{\beta}{\beta_2}. \tag{8.23}$$

From the formula (8.8) it follows that

$$\varphi = \sqrt{2\Delta}. \tag{8.24}$$

Using the expression (8.22), we get

$$\kappa = \sqrt{8\Delta} - 1.$$

From the linearized equation for χ_1 we find

$$\chi_1 = \sqrt{2\Delta}\left(\sqrt{2\Delta} - 1\right). \tag{8.25}$$

Then

$$\theta_1 = 2\Delta. \tag{8.26}$$

Thus, when $\beta \to \beta_2$ the values of the angles θ_j for the volume differ from zero by an order of magnitude $\sqrt{\Delta}$, while on the surface this difference is an order of magnitude Δ, which is to be expected, since the magnetic moment of the upper atomic plane makes a smaller angle with the direction of the magnetic field than in the volume.

The radius of correlation $r_c = \frac{d}{\sqrt{3\Delta}} \propto \Delta^{-\frac{1}{2}}$ as it follows from the mean-field theory for the second order phase transitions.

Thus, additional (in comparison with the volume) distortions of the antiferromagnetic order parameter near the uncompensated surface of the antiferromagnet caused by the magnetic field occur in the field of the surface spin-flop transition β_s and are present up to the field of the spin-flip transition to the ferromagnetic phase [3].

8.3 Size Effects in a Plane-parallel Layer of an Antiferromagnet with Smooth Surfaces

$\alpha \ll 1$ the case of an uncompensated surface of an antiferromagnet, the radius of correlation of the order parameter r_c in the fields $\beta \sim \beta_1$ is much greater than the interatomic distance, so in this field area, as well as near the spin-flip transition field, when r_c becomes of the order of the thickness of the antiferromagnetic layer or exceeds it, noticeable size effects can occur.

As mentioned above, a system consisting of alternating ferromagnetic and non-magnetic metal layers, with an antiferromagnetic sign of exchange between the nearest ferromagnetic layers, behaves in a magnetic field similar to the considered layer of an antiferromagnet.

The theoretical study of such multilayer structures was carried out by numerical methods in many works but attempted to consider the size effects were practically not made analytically.

The changes in formulas (8.13)–(8.16) when describing an antiferromagnet of finite thickness are that now the number of terms in the sum j is finite and is equal M to the number of atomic planes in the layer, and the term $\cos(\theta_j - \theta_{j+1})$ in the right part of Eq. (8.13) and the term $\sin(\theta_j - \theta_{j+1})$ in the left part of Eq. (8.16) must be multiplied by a bracket $(1 - \delta_{M,j})$.

The behavior of layers with an even and odd number of atomic planes differs fundamentally. On the example of multilayer structures with $M = 15$ and $M = 16$, such a difference was first discovered in [R. W. Wang et al. *Phys. Rev. Lett.* **72,** 920 (1994)].

8.3.1 Even Number of Atomic Planes

In such a layer, the surface atomic planes belong to different sublattices of the antiferromagnet. In magnetic fields $\beta < \beta_s$, the layer is in the collinear phase. The spin-flop transition occurs in the field $\beta = \beta_s$, and as a result, a 180° Bloch domain wall appears in the center of the layer. If the correlation radius $r_c(\beta_s)$ given by the formula (8.20) is less than the layer thickness a, then the wall occupies only the central part of the layer (Curve 1 in Fig. 8.3). Due to the appearance of the wall, the magnetic moments of the surface atomic planes are directed along the magnetic field.

If the thickness $a < r_s \equiv r_c(\beta_s) = d\beta_s^{-1} \approx \frac{d}{\sqrt{2\alpha}}$, then only the central part of the domain wall is placed in the layer, in which the angle of rotation of the antiferromagnetism vector changes linearly with the distance.

As the effective field grows and the value β approaches the value β_1, the value of the correlation radius diverges in a root-like manner. In this case, at $a \gg r_s$, the domain wall, expanding, captures the entire thickness of the layer in the field β_a, that can be found from the condition

$$2r_c(\beta_a) \approx a. \tag{8.27}$$

With further growth r_c, the width of the domain wall ceases to grow, and the antiferromagnetism vector turns uniformly at an angle π from one surface of the layer to the other, so that the magnetizations of the edge atomic planes are directed along the external field (line 2 in Fig. 8.3).

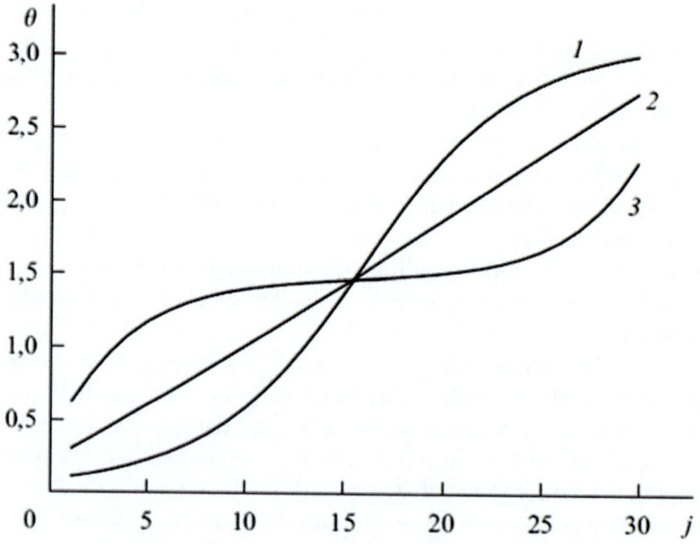

Fig. 8.3 Domain walls in a layer with an even number of atomic planes in the fields $\beta_s < \beta < \beta_1(1)$; $\beta \approx \beta_1(2)$; $\beta > \beta_1(3)$

In exchange fields that exceed the field β_1, the value r_c begins to decrease, and the 180° domain wall in the center of the layer is transformed into two almost 90° exchange spirals near the surfaces (Curve 3 in Fig. 8.3).

In this case, the center of the layer is in the spin-flop phase. As the field increases, the magnitude r_c decreases, the size effects become insignificant, and the behavior of the exchange spiral near each layer surface is described independently of the other. The patterns of this behavior are discussed in Sect. 8.2.

As the field approaches the field of the volumetric spin-flip transition β_2, the value r_c diverges, and the size effect becomes significant again.

As shown by the simulation, the spin-flip transition in a layer of finite thickness is a second order phase transition and occurs in the field $\beta_2^* < \beta_2$ (Fig. 8.4).

The formula can approximate the resulting dependence

$$\beta_2^* = \beta_2\left[1 - \frac{1}{8}\left(\frac{4,5}{M}\right)^2\right]. \tag{8.28}$$

The phase diagram of a layer with an even number of atomic planes in the "magnetic field-layer thickness" variables is shown in Fig. 8.5.

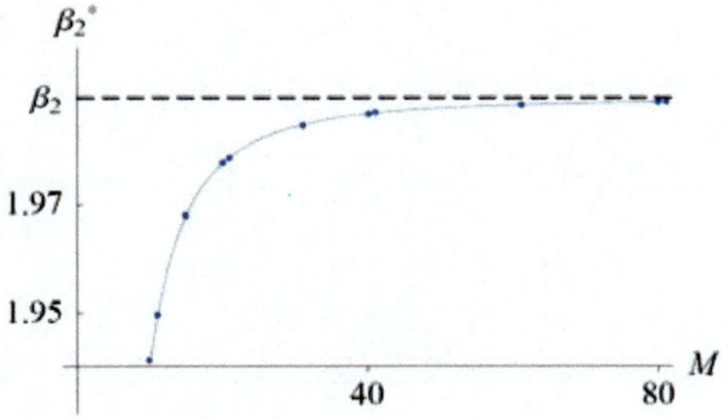

Fig. 8.4 Dependence of the spin-flip transition field on the number of atomic planes in the layer

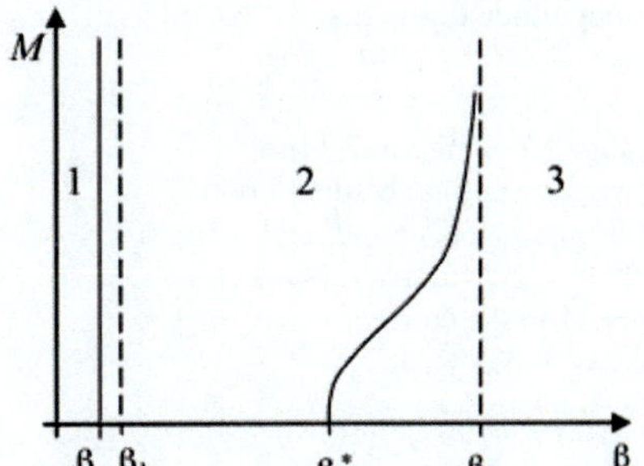

Fig. 8.5 Phase diagram "magnetic field-layer thickness" for a layer with an even number of atomic planes: collinear antiferromagnetic phase **(1)**, non-collinear phase **(2)**, ferromagnetic phase **(3)**. Solid lines correspond to the lines of phase transitions

8.3.2 *Odd Number of Atomic Planes*

In this case, both surface atomic planes belong to the same sublattice of the antiferromagnet. We will proceed from the assumption that the total magnetic moment and consequently the magnetic moments of these planes are directed along the external magnetic field.

As shown in Sect. 8.1, the spin-flop transition in the surface layer is suppressed in this configuration. Therefore, in a layer of finite thickness, the spin-flop transition occurs in a field $\beta_1^* > \beta_1$, and, unlike an infinite sample and a layer with an even number of atomic planes, this is a second order phase transition. The rotation angle of the magnetizations changes continuously with the field, and the maximum angle of rotation occurs in the center of the layer (Figs. 8.6, 8.7).

The expression approximates the calculated dependence $\beta_1^*(M)$ shown in Fig. 8.8

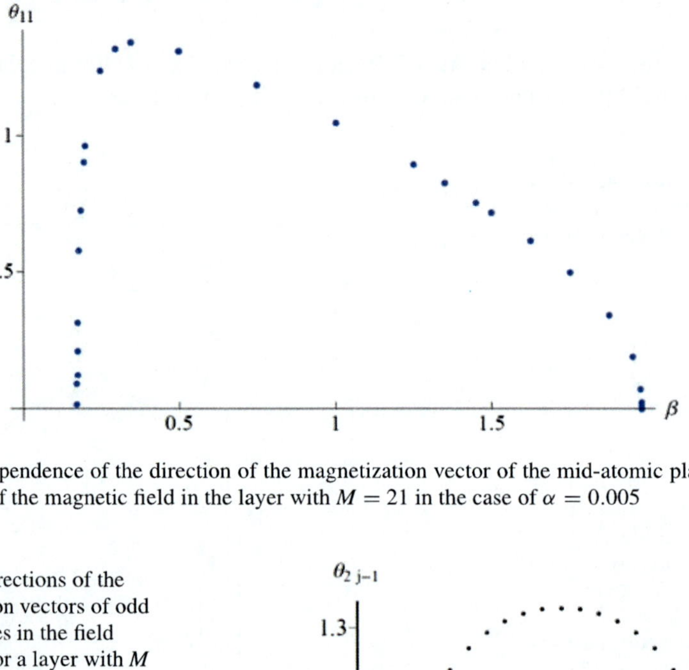

Fig. 8.6 Dependence of the direction of the magnetization vector of the mid-atomic plane on the magnitude of the magnetic field in the layer with $M = 21$ in the case of $\alpha = 0.005$

Fig. 8.7 Directions of the magnetization vectors of odd atomic planes in the field $\beta = 0, 2\ 2$ for a layer with $M = 41$ in the case of $\alpha = 0, 005$

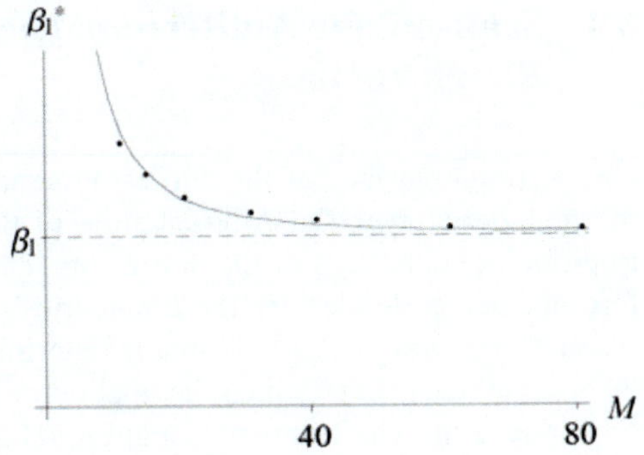

Fig. 8.8 Dependence of the spin-flop transition field on the number of atomic planes in a layer with an odd number of atomic planes

$$\left(\beta_1^*\right)^2 = \beta_1^2 + \left(\frac{4}{M}\right)^2. \tag{8.29}$$

The behavior of a layer with odd M margins $\beta > \beta_1^*$ is similar to the case of an even M layer.

The phase diagram of a layer with an odd number of atomic planes is shown in Fig. 8.9.

Thus, the nature of the spin-flop transition in a layer is determined by the parity of the number of atomic planes in it. If for the even M spin-flop transition is a first order phase transition and always occurs in the field of the surface spin-flop transition, then for the odd M one has a second order phase transition, which occurs in a field that exceeds the field of the volume spin-flop transition and depends on the magnitude M.

The spin-flip transition in a layer of finite thickness occurs in a field which is less than the field of the volume spin-flip transition.

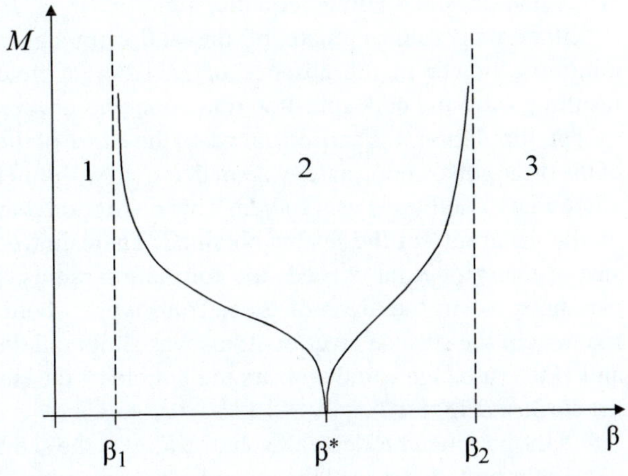

Fig. 8.9 Phase diagram "magnetic field-layer thickness" for a layer with an odd number of atomic planes: collinear antiferromagnetic phase **(1)**, non-collinear phase **(2)**, ferromagnetic phase **(3)**. Solid lines correspond to the lines of phase transitions

8.4 Semi-infinite Antiferromagnet with Uncompensated Rough Surface

On the rough surface of the antiferromagnet, there are atomic steps with a height of one atomic layer. On different sides of the step, the upper atomic layer belongs to different sublattices of the antiferromagnet. Thus, the entire surface of the antiferromagnet is divided by the boundaries of the steps into areas of the first and second type, and the type of area is determined by the number of the sublattice of the antiferromagnet to which the upper atomic layer belongs.

Suppose the characteristic longitudinal size of the steps R is large enough. In that case, when a magnetic field is applied parallel to the magnetization of the first sublattice, the value of the field β satisfies the inequality $\beta_s < \beta < \beta_1$, the magnetic structure of the first-type areas will remain unchanged, and in the second-type areas, a surface spin-flop transition will occur. As a result, the surface will break into domains [4].

The type of surface distortions of the magnetic structure significantly depends on the ratio between the characteristic width of the steps R and the radius of correlation of the order parameter r_c. A strong dependence r_c on the magnitude of the magnetic field β and, in particular, the divergence r_c at $\beta \to \beta_1$, leads to a rather complex phase diagram of the distortions of the magnetic structure in the variables (β, R). Consider a tetragonal volume-centered lattice of spins whose direction is given by a unit vector $S_{n,i,j}$, with the index n numbering the planes starting from the uppermost one present in the sample with a rough surface, and the indices i and j specifying the position of the spins in the plane where they form a rectangular lattice. The spin value was assumed to be constant and included in the corresponding interaction constants.

Along with the Heisenberg exchange interaction (2.2) and the uniaxial anisotropy power (4.1), the Zeeman power (6.7) was taken into account. The equilibrium distribution of spins was found by modeling their behavior based on the system of Landau–Lifshitz–Hilbert equations.

Either the collinear state of the antiferromagnet, which corresponds to the minimum power in the absence of an external field, or the homogeneous state resulting from the bulk spin-flop transition was chosen as the initial state.

The simulation was carried out when the edges of the steps are parallel to the axis y of the orthogonal coordinate system; that is, a two-dimensional problem with periodic x boundary conditions was solved. This restriction does not have a qualitative effect on the generality of the results obtained. They showed that when the characteristic size of the steps R far exceeds the correlation radius of the antiferromagnetic order parameter r_c, in the areas of the second type, where in weak fields the magnetic moment of the surface layer of atoms was antiparallel to the external magnetic field, and 180° exchange spiral appears in the field of the surface spin-flop transition near the surface (Fig. 8.10).

As the parameter R decreases, but at $R > r_c$ the same time, the spin-flop transition field in the bounded area of the second type shifts towards more strong magnetic fields (Fig. 8.11). If $R < r_c$ the domain wall in a separate area is not formed at all.

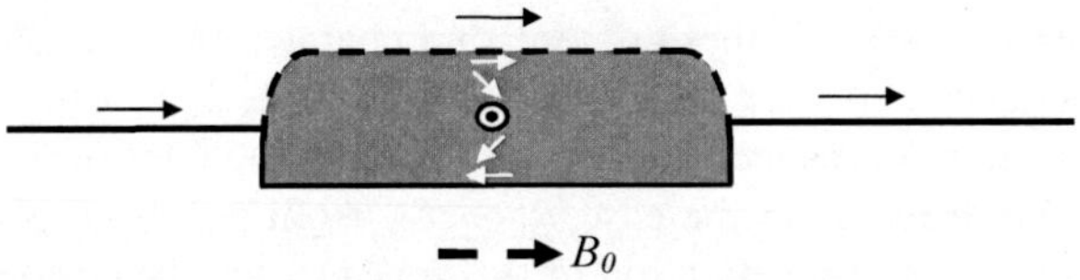

Fig. 8.10 Exchange spiral in the area of the second type. Solid arrows indicate the local direction of the antiferromagnetism vector. Its rotation occurs in a plane parallel to the surface. The dotted arrow shows the direction of the magnetic field

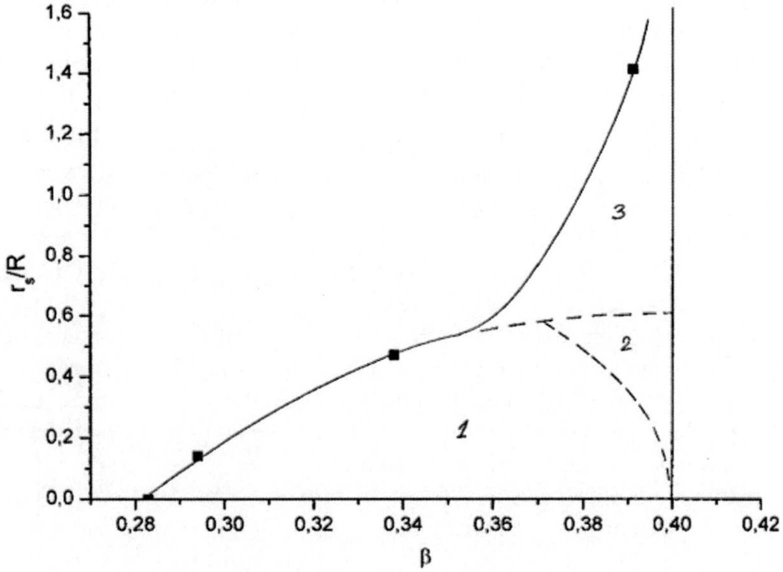

Fig. 8.11 Phase diagram "magnetic field-roughness" in the field $\beta \leq \beta_1$ area in the case of $\alpha = 0.04$. A solid line is a line of the first order phase transitions from a collinear phase to a non-collinear one. The numbers indicate the areas of existence of domains **(1)**, vortices **(2)**, and weak distortions **(3)**

When the external field approaches the value of the field of the bulk spin-flop transition β_1, $r_c \to \infty$, therefore, even in the case of the domain phase, the ratio $R > r_c$ is violated as the magnetic field increases. For the field region, in which $r_c \gg R$ the problem under consideration is reduced to the problem of distortions of the order parameters at the rough interface of a two-layer ferromagnet–antiferromagnet system, assuming that the ferromagnet is absolutely rigid concerning magnetic distortions, and the interlayer exchange interaction be replaced by the action of an external magnetic field β.

As the calculation shows, near the surface of the antiferromagnet $|z| < R$, static spin vortices have formed that capture the area, the boundaries of which on the surface of the antiferromagnet coincide with the edges of the atomic steps. Moreover, along with the characteristic size R, there is a second characteristic size δ_0–the width of the surface area near the atomic step, in which the value $\theta_{n,i,j}$ differs significantly from its optimal value (0, respectively, and π on different sides of the step).

We estimate the value δ_0 in the case $R \gg \delta_0$ from power considerations. The fact that the direction of magnetization of the upper atomic layer does not coincide with the direction of the magnetic field in the area of width δ_0 leads to a loss in the Zeeman power (here and further in units $\tilde{z}|J|$) of the order $\beta\frac{\delta_0}{S_0}$ of one unit of the length of the atomic step, where S_0 is the area of a rectangular cell in the atomic plane. The loss in anisotropy power $\alpha\frac{\delta_0}{S_0}$ is negligible in comparison with it.

In the area $\delta_0 < \rho < R$ where ρ is the shortest distance from the point to the edge of the step on the surface of the antiferromagnet, $|\nabla\theta|$ is inversely proportional to ρ. In this area, the contribution of the distortions of the antiferromagnetic order parameter to the exchange power is, per unit length of the atomic step, the value of the order $b_y^{-1}\ln\left(\frac{R}{\delta_0}\right)$, where b_y is the parameter of the crystal lattice along the edge of the nuclear stage. Minimizing the total power by the parameter δ_0, we find $\delta_0 \sim \frac{b_x}{\beta}$ (b_x–the parameter of the crystal lattice along *the x-axis*), and in the area, $\beta \sim \beta_1$ the value of $\delta_0 \sim r_s$.

In the opposite case $R \ll \delta_0$, the reversal of the magnetization does not have time to occur. That is the area of weak distortion.

As already noted, the steps divide the entire surface of the antiferromagnet into two types of regions with total areas σ_1 and σ_2, respectively. If the average value θ in the region $R \ll |z| \ll r_c$ is equal ψ, then in the vortex occupying the region of the first type, the angle changes θ from zero to ψ, while in the vortices in the regions of the second type, the angle θ changes from ψ to π.

By analogy with Slonczewski's "magnetic proximity" model, the total power of vortices can be written as (4.20) with

$$C_k \propto {}^{\sigma_k}/_{R}. \tag{8.30}$$

In the case $\sigma_1 = \sigma_2$ of the minimum power of vortices corresponds to $\psi = \frac{\pi}{2}$.

Thus, to reduce the power of vortices, it is necessary to create 90-degree exchange spiral near the surface of the antiferromagnet (Fig. 8.12). Its appearance is associated with additional surface power w, which, per unit cell on the surface of the crystal, is on the order of magnitude equal to [5]

$$w \sim \sqrt{\beta_1^2 - \beta^2}, \tag{8.31}$$

and turns to zero at $\beta \to \beta_1$. For the considered field range, in the case where $\delta_0 \ll R \ll r_c$ the gain in the power of vortices is much greater than the loss in power due to the formation of an exchange spiral.

When $r_s \ll R$ the transition from the domain phase to the vortex phase occurs continuously in the field β_R for which $r_c(\beta_R) \sim R$, that is, when

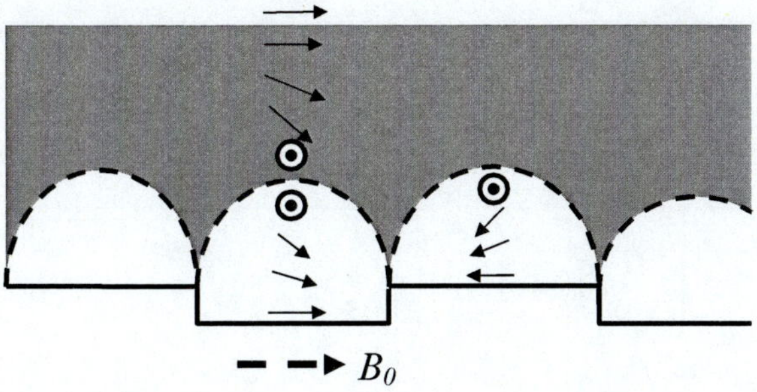

Fig. 8.12 Static spin vortices and a 90° exchange spiral. The area of existence of the spiral is highlighted in gray. Solid arrows indicate the local direction of the antiferromagnetism vector. Its rotation occurs in a plane parallel to the surface. The dashed arrow shows the direction of the magnetic field

$$\tau_R \equiv 1 - \frac{\beta_R}{\beta_1} \sim \frac{r_s^2}{R^2} \tag{8.32}$$

The "magnetic field-roughness" phase diagram for $\beta < \beta_1$ is shown in Fig. 8.11. For $R \ll r_S$ the phase transition from the collinear state occurs immediately in the phase with weak distortions.

As the discrete model simulation shows, the transition from a collinear phase to a phase with weak distortions and a 90° exchange helix is a first order surface phase transition occurring in a field $\widetilde{\beta_R}$ whose magnitude can be estimated from simple power considerations. Corresponding value

$$\widetilde{\tau_R} \equiv 1 - \frac{\widetilde{\beta_R}}{\beta_1} \sim \left(\frac{R}{r_S}\right)^2 \ll 1. \tag{8.33}$$

In the field $\beta > \beta_1$, when the spin-flop transition in the antiferromagnet volume has already occurred, the behavior of the surface distortions is almost the same for all values R. Since $r_c \to \infty$ at $\beta \to \beta_1 + 0$, near the field β_1, the system is in the vortex phase or the phase of weak distortions, and the surface distortions penetrate the antiferromagnet to a depth of about R. In the case $\sigma_2 = \sigma_1$, as a result of a volume spin-flop transition, the surface exchange spiral disappears.

As the field grows, r_c decreases according to the formula (8.21), and the value δ_0 changes as β^{-1}. At $\beta \geq 1$ and not too close to the collapse field of sublattices β_2, these are quantities of the same order and are comparable to the lattice constant.

In the field β'_R, for which $r_c(\beta'_R) = R \gg \delta_0(\beta'_R)$ there is a smooth transition to the domain phase, in which the regions of the two types are formed: almost 90° exchange spirals with opposite direction of rotation of the antiferromagnetic order parameter (from $\theta = 0$ on the surface to $\theta = 90°$ in the volume in the region of the first type and from $\theta = 180°$ at the surface to $\theta = 90°$ in the volume in the region of the second type). For the field β'_R, we get the estimate [6]

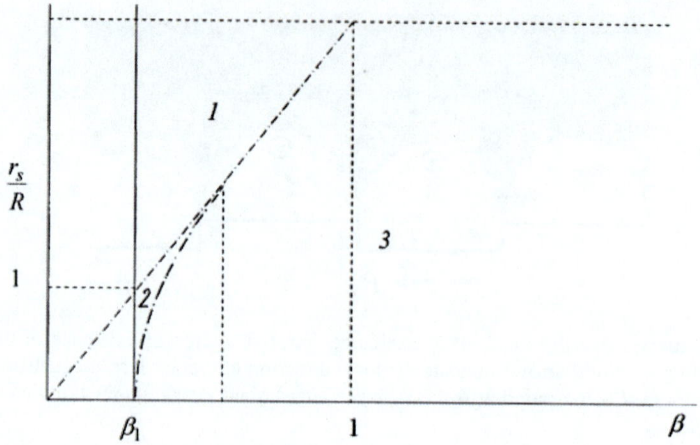

Fig. 8.13 Phase diagram "magnetic field-roughness" in the field range in the case of $\beta > \beta_1$. The numbers indicate the areas of existence of weak distortions **(1)**, vortices **(2)**, and domains **(3)**

$$\tau'_R = \frac{\beta'_R}{\beta_1} - 1 \sim \frac{r_s^2}{R^2}. \tag{8.34}$$

As the field increases, the magnetization of the upper layer of atoms rotates, deviating from the direction of the magnetic field and approaching the direction of the magnetization of the corresponding sublattice in the volume. The rotation angle in the exchange spiral decreases, and its transverse size decreases up to the field $\beta^* = \frac{\beta_2}{\sqrt{2}}$.

At $\beta_2 > \beta > \beta^*$ the same time, the turn of the antiferromagnetic order parameter in regions of different types, in contrast to the range of fields $\beta_1 < \beta < \beta^*$, has one sign, so the difference between the regions is leveled out, and the further evolution of surface distortions is similar to the case of a smooth surface, studied in Sect. 8.2.

If $R < \delta_0(\beta)$, then only weak surface distortions are observed. The phase diagram in the field $\beta_2 > \beta > \beta_1$ range is shown in Fig. 8.13.

8.5 Antiferromagnet Layer with an Uncompensated Rough Surfaces

Antiferromagnetic layers of nanometer thickness are widely used in modern magnetoelectronics. Therefore, the study of their characteristics is of undoubted interest.

The antiferromagnetic layer can be mentally divided into areas whose boundaries are perpendicular to the plane of the layer. The boundary plane passes through the edge of the atomic step on one of the layer surfaces. The magnetization behavior in each area is determined by the number of atomic planes belonging to each of the two sublattices of the antiferromagnet. Inside the area, this number remains unchanged. A combination of two parameters determines the type of area: the parity of the number of atomic planes (*E* (even)–even, *O* (odd) is an odd number) and the number of the sublattice of the antiferromagnet to which the upper atomic plane (*A* or *B*) belongs. There are four possible combinations, and therefore four types of areas (*EA*, *EB*, *OA*, and *OB*).

As shown in Sect. 8.3, in the case of a layer with smooth boundaries, the character of the spin-flop transition depends significantly on the parity of the number of atomic planes [2]:

- in a layer with an even number of atomic planes, a first order spin-flop transition occurs in the field of the surface spin-flop transition. The result is a state with the domain wall centered in the middle of the layer;
- in a layer with an odd number of atomic planes, a second order spin-flop transition occurs in a field that exceeds the field of the bulk spin-flop transition and depends on the thickness of the layer.

If the characteristic distance between the steps R is large enough, then in the magnetic field the antiferromagnetic layer is divided into domains whose boundaries coincide with the boundaries of the areas, that is, perpendicular to the plane of the layer and intersect the edge of the atomic step on one of its surfaces. The type of the corresponding area dictates the behavior of the magnetization in each domain.

A layer cut perpendicular to the direction [100] of the tetragonal volume-centered lattice was selected to find the nature of the magnetic structure distortions in the entire range of magnetic fields and values of the magnitude R within the framework of the model described in the previous section.

The behavior of thin ($a \ll r_s$) and thick ($a \gg r_s$) layers of antiferromagnet in a magnetic field $\beta \leq \beta_1$ differed significantly. Therefore, we will describe these cases sequentially.

8.5.1 Thin Layer

(a) $R \gg r_s$

We study the distribution of spins in regions of all types, assuming that the size of the domain R exceeds other characteristic lengths of the problem.

Regions *OA* and *OB* have a magnetic moment equal to the magnetic moment of one uncompensated atomic plane. Therefore, a second order spin-flop transition in the area *OA* will occur in the field β_1^* (formula (8.29)). In the case of a thin layer, we obtain from (8.29) [4]

$$\beta_1^* \approx \frac{4}{M} \approx \frac{4\beta_s r_s}{a} \gg \beta_1. \quad (8.35)$$

In the *OB* area, the magnetic moment when the field is turned on is directed against it. In an *OB*-type layer with smooth boundaries, the direction of the antiferromagnetism vector is reversed in an arbitrarily weak field. Therefore, this initial orientation was not investigated in Sect. 8.3. In the case of an area of finite dimensions, the reorientation of the antiferromagnetism vector is accompanied by a change in its type from *OB* to *OA* and the appearance of a 180° domain wall perpendicular to the layer surface along the perimeter of the area.

The characteristic field of this phase transition can be found by equating the gain in the Zeeman power to the power of the resulting domain wall. The gain in the Zeeman power is a value of the order β of one cell in the plane of the layer (here and further on, the power is given in units $\tilde{z}|J|$). The surface power of the domain wall in an antiferromagnet is of the order of magnitude $\sqrt{\alpha}$ per unit cell area, and its thickness is of the order s of magnitude r_s. Since the area of the region on the surface of the layer $\sim R^2$, and the area of the domain wall $\sim Ra$, the characteristic reorientation field is equal to

$$\beta_{\text{reor}} \approx \frac{a\sqrt{\alpha}}{R} \approx \frac{a}{R}\beta_1 \ll \beta_1. \tag{8.36}$$

As will be shown below, the resulting domain wall is almost completely located in a region with an even number of atomic planes, since the formation of a domain wall in such a region is not accompanied by a noticeable decrease in the Zeeman power.

To describe the domain walls perpendicular to the layer, we use the continuum approximation.

Let's replace the variables $\tilde{\theta}_{2n-1,i,j} = \theta_{2n-1,i,j}, \theta_{2n,i,j} = \pi + \tilde{\theta}_{2n,i,j}$. In fields much smaller than the spin-flip transition field β_2

$$\left|\tilde{\theta}_1(x, y, z) - \tilde{\theta}_2(x, y, z)\right| \ll 1, \tag{8.37}$$

where the functions $\tilde{\theta}_1(x, y, z)$ and $\tilde{\theta}_2(x, y, z)$ describe the direction of the spins belonging to the first and second sublattices of the antiferromagnet, respectively.

Minimizing the total power W in terms of variables $\tilde{\theta}_{n,i,j}$ and passing to the continuum representation in the resulting equations, we obtain, taking into account (8.37), in the volume of the antiferromagnet

$$\begin{cases} \tilde{\theta}_2 - \tilde{\theta}_1 + \frac{1}{8}\Delta\tilde{\theta}_2 = \alpha \sin 2\tilde{\theta}_1 + \beta \sin \tilde{\theta}_1 \\ \tilde{\theta}_1 - \tilde{\theta}_2 + \frac{1}{8}\Delta\tilde{\theta}_1 = \alpha \sin 2\tilde{\theta}_2 - \beta \sin \tilde{\theta}_2, \end{cases} \tag{8.38}$$

where Δ is the three-dimensional Laplacian in dimensionless variables $\zeta = \frac{x}{b_x}, \eta = \frac{y}{b_y}, \varsigma = \frac{z}{b_z}$, $b_z = b_y$ the size of the tetragonal cell of the crystal along *the z-axis.*

Entering variables

$\theta = \frac{\tilde{\theta}_1 + \tilde{\theta}_2}{2}$ and $\varphi = \tilde{\theta}_2 - \tilde{\theta}_1 \ll 1$, we find by adding and subtracting the Eqs. (8.38):

$$\frac{1}{8}\Delta\theta = \alpha \sin 2\theta - \frac{1}{2}\beta\varphi \cos\theta, \tag{8.39}$$

$$\varphi + \frac{1}{16}\Delta\varphi = \beta \sin\theta. \tag{8.40}$$

In (8.39) and (8.40), we considered that $\alpha,\varphi \ll 1$ and neglected the terms containing the second and higher powers of these parameters.

If the sizes of the magnetic inhomogeneities are much larger than the atomic sizes, it can be neglected $\Delta\varphi$ in Eq. (8.40). Then it takes the form

$$\varphi = \beta \sin\theta. \tag{8.41}$$

Substituting the value φ in the Eq. (8.39), we get

$$\Delta\theta = 2\left(\beta - 1^2 - \beta^2\right) \sin 2\theta. \tag{8.42}$$

Let us now consider the situation at the boundary of the antiferromagnet. If the upper one is the atomic plane belonging to the first sublattice, then as a result of the transition to the continuum representation in the equation for the spin belonging to this plane, we have,

$$\tilde{\theta}_2 - \overline{\theta}_1 - \frac{1}{2}\frac{\partial\tilde{\theta}_2}{\partial\varsigma} + \frac{1}{8}\Delta_{\xi,n}\tilde{\theta}_2 = 2\alpha \sin 2\tilde{\theta}_1 + 2\beta \sin\tilde{\theta}_1, \tag{8.43}$$

where $\Delta_{\xi,n}$ is the two-dimensional Laplacian in the plane of the layer. If the upper plane belongs to the second sublattice, then the boundary condition takes the form

$$\tilde{\theta}_1 - \tilde{\theta}_2 + \frac{1}{2}\frac{\partial\tilde{\theta}_1}{\partial\varsigma} + \frac{1}{8}\Delta_{\xi,n}\tilde{\theta}_1 = 2\alpha \sin 2\tilde{\theta}_2 - 2B \sin\tilde{\theta}_2. \tag{8.44}$$

Using the continuity of the function $\varphi(x, y, z)$, after passing to the variables θ and φ and substituting (8.41) into the Eqs. (8.43), (8.44) we find

$$-\frac{1}{2}\frac{\partial\theta}{\partial\varsigma} + \frac{1}{8}\Delta_{\xi,n}\theta = \pm\beta \sin\theta, \tag{8.45}$$

where the plus or minus signs on the right side correspond to the cases where the upper atomic plane belongs to the first or second sublattice, respectively. On the opposite surface of the layer, it is necessary to change the sign before the derivative $\frac{\partial\theta}{\partial\varsigma}$ in Eq. (8.45).

Using the inequality $a \ll r_s$, we integrate Eq. (8.42) over the thickness of the layer, and all the terms, with the exception $\frac{\partial^2\theta}{\partial\varsigma^2}$, are represented as values averaged over the layer, and $\frac{\partial^2\theta}{\partial\varsigma^2}$ we find the integral form using the Newton–Leibniz formula, using $\frac{\partial\theta}{\partial\varsigma}$ the boundary conditions (8.45) to find the values at the boundaries of the layer. As a result, assuming that the number of planes in the layer $M \gg 1$, we get in the case of odd M

$$\Delta_{\xi,n}\theta = \pm\frac{8^{\beta}}{M}\sin\theta + 2\left(\beta_1^2 - \beta^2\right)\sin 2\theta. \tag{8.46}$$

The plus sign corresponds to the type *OA*, and the minus sign is an area of the type *OB*.

For even-numbered areas M

$$\begin{aligned}
&\left.\frac{\partial\theta}{\partial\varsigma}\right|_{\varsigma=\frac{M}{4}} - \left.\frac{\partial\theta}{\partial\varsigma}\right|_{\varsigma=-\frac{M}{4}} \\
&\approx \mp 2\beta\left(\sin\theta|_{\varsigma=M/4} - \sin\theta|_{\varsigma=-M/4}\right) \\
&\approx \mp\beta M\cos\theta\frac{\partial\theta}{\partial\varsigma} \\
&\approx \beta^2 M\sin 2\theta.
\end{aligned} \tag{8.47}$$

As a result, after averaging over the layer thickness for the areas *EA* and *EB* we find

$$\Delta_{\xi,n}\theta = 2\left(\beta_1^2 - 2\beta^2\right)\sin 2\theta. \tag{8.48}$$

When considering a flat domain wall parallel to the plane *yz*, the problem becomes one-dimensional.

It is easy to see that already in relatively weak fields $\beta_1 \gg \beta > \left(a/r_s\right)\beta_2$, the domain wall in the areas *OA* and *OB* has a width of

$$\lambda_{\text{odd}} = b_x\left(\frac{M}{8\beta}\right)^{\frac{1}{2}}, \tag{8.49}$$

much smaller than the value r_c and than the wall width λ_{even} in an area with an even number of atomic planes

$$\lambda_{\text{even}} = \frac{b_x}{2\sqrt{\left|\beta_1^2 - 2\beta^2\right|}}. \tag{8.50}$$

Accordingly, the surface power of such a wall is significantly higher, which is why the domain wall perpendicular to the layer surface, which occurs in the reorientation field at the boundary of the *OB* area, is almost entirely located in the area with an even number of atomic planes. In the case of parallel edges of the nuclear stages, it is located in the middle of the area with the type *EA(EB)* and separating the areas of type *OA* and *OB*.

The width of the domain wall increases with the growth of the magnetic field, diverging root-wise in the field of the surface spin-flop transition (and not the bulk one, as the radius of the correlation of the order parameter). At $\beta \to \beta_2$ the same time, the domain wall occupies the entire area of type *EA(EB)*, and in fields that exceed the field of the surface spin-flop transition, it smoothly transforms into two 90° domain walls located at the boundaries of the area (similar to how it is depicted in Fig. 8.3). In the center of the area, there is a spin-flop state discussed in Sect. 8.3.

If the area *of EA(EB)* lies between two areas *of the OB*, then in the field of reorientation in it, as well as in neighboring areas, the antiferromagnetic parameter turns by about 180°, the domain wall does not form, and the area of type *EA* becomes an area of type *EB* (and vice versa).

Spin-flop transition in the area of the type *EA (EB)*, separating the area of the type *OA*, is a first order phase transition. As a result of interaction with neighboring areas, the antiferromagnetism vector in the center of the area abruptly changes by an amount less than $\pi/2$, and domain walls appear at the areas' boundaries. As the magnetic field increases, the antiferromagnetism vector is oriented perpendicular to the field. The field of this phase transition β_s^* exceeds the value β_s and can be estimated from the condition $2\lambda(\beta_s^*) \approx R$. Using the expression (8.50), we find [5]

$$\beta_s^* \approx \left[\beta_s^2 + \frac{1}{2}\left(\frac{b_x}{R}\right)^2\right]^{1/2}. \tag{8.51}$$

Thus, in the case of a weak roughness of the layer, when the width of the steps $R \gg r_s$, as the magnetic field in the layer increases, there will be sequential: a reorientation transition in *OB*-type areas, a spin-flop transition in EA-type areas (EB), and the spin-flop transition in *OA*-type areas, respectively. The mutual influence of different types of areas, the finiteness, and the difference in their sizes leads to some smearing of these phase transitions.

(b) $R \ll r_s$

In strong roughness, the reorientation of individual areas of the *OB* type is impossible, and a spin-flop transition is realized in the entire volume of the layer.

We find the characteristic field of the "collective" spin-flop transition at $R \ll r_s$. We consider the case when the total areas σ_{OA} and $\sigma_{\mathrm{O}\beta}$ the surface of the layer occupied by the type *OA* and *OB* areas are the same. Otherwise, the system's behavior is determined by the dominant type of areas, and the consideration is similar to the one conducted above. It is only necessary to consider in the power balance a factor equal to the ratio $(\sigma_{\mathrm{OA}} - \sigma_{\mathrm{O}\beta}) / \sigma$ (σ-the total surface area of the layer).

Assume that the antiferromagnetism vector averaged over the volume of the layer makes an angle θ with an easy axis. In areas of the type OA and OB vector of antiferromagnetism is additionally rotated in different directions by an angle $\chi \ll 1$ so that the magnetization of the extra (unpaired) atomic plane makes an angle as small as possible with the magnetic field. The corresponding gain in the Zeeman power will be of an order of magnitude $\beta X b_x \sin\theta / a$ per cell, and the loss in the exchange power at $R \gg a$ will be an order of magnitude $x^2 b_x^2 / R^2$. Minimizing the total power of X, we find that $x \approx \beta R^2 \sin\theta / a b_x$ the gain in total power is of the order of magnitude $\beta^2 R^2 \sin^2\theta / a^2$ per cell. In the spin-flop transition field, it must compensate for the loss in power by overturning the sublattices, equal to $(\beta_1^2 - \beta^2)\sin^2\theta$. As a result, we get the spin-flop transition field β_{s_f}

$$\beta_{sf} \approx \beta_1 \frac{\alpha}{R} \ll \beta_1. \tag{8.52}$$

It is easy to check that in the entire area of strong roughness $x(B_{s_f}) \ll 1$ and $\lambda_{\mathrm{odd}}(\beta_{sf}) \gg R$.

If $R \ll a$, then the inhomogeneity of the antiferromagnetic order parameter captures only areas with the width of order of R near each of the layer surfaces, a loss in the exchange power of order $x^2 b_x^2 / R\alpha$, $x \approx \beta R b_x^{-1} \sin\theta$ and

$$\tau_{s_f} \equiv 1 - \frac{B_{sf}}{\beta_1} \approx \beta_1 \frac{R}{\alpha} \propto R,\ \tau_{sf} \ll 1. \tag{8.53}$$

With the growth of the field, the value X increases and $\beta \approx \beta_0$ becomes of the order of one. In the case $\beta_0 \ll \beta_1^*$ that is equivalent to the condition $R \gg a$, in high fields β_0, the behavior of areas of different types becomes "individual" and coincides with the one considered in the previous case. In the opposite limit $R \ll a$, the system remains in the "collective" spin-flop phase.

In fields $\beta \gg \beta_1^*$, but smaller than the spin-flip transition field, when the correlation radius of the antiferromagnetic order parameter $r_c \ll a$, the volume of the layer is in the spin-flop phase, and distortions of the order parameter near one of the layer surfaces do not affect those of the other surface. This mode is discussed in Sect. 8.4. Near the spin-flip transition field, the roughness does not play a significant role, and the dimensional effect is $\beta \to \beta_2$ similar to that described in Sect. 8.3.

The phase diagram "magnetic field-roughness" of a thin layer of antiferromagnet is shown in Fig. 8.14.

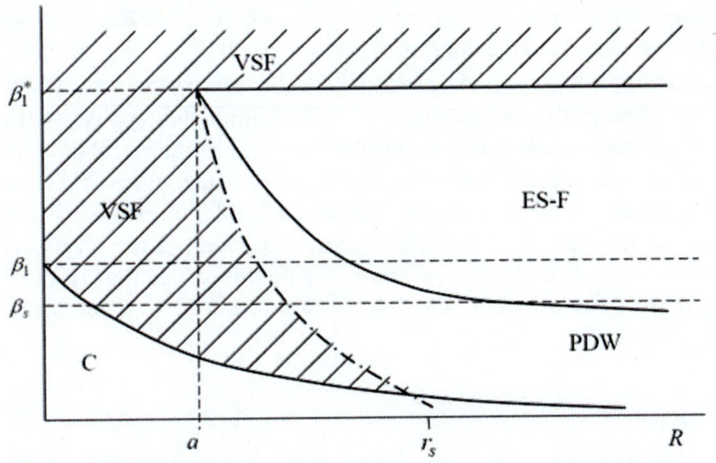

Fig. 8.14 "Magnetic field-roughness" phase diagram for a thin layer: C-collinear phase, PDW-phase with domain walls perpendicular to the surface of the layer, resulting from the reorientation of areas of the type *OB*, ES-F-the phase in which the spin-flop transition occurred only in areas of the type *EB* and *EA*, VS-F-volume spin-flop phase. Solid lines indicate the lines of phase transitions

8.5.2 Thick Layer

(a) $a \gg R \gg r_s$

In this case, the surface domain partitioning occurs independently near each of the two-layer boundaries in the field β_s. The initial stage of changes like the magnetic structure distortions as the magnetic field increases is described in Sect. 8.4. As the correlation radius increases r_c, its value is compared with R in the field β_R specified by the formula (8.32).

At $\beta > \beta_R$ a system of static vortices is formed near each surface of the layer, penetrating deep into the layer at a distance of about R, and an exchange spiral with a width parallel to the surface of 90° r_c is formed.

With the growth of the field, the exchange spirals, due to divergence r_c, capture an increasing share of the layer, and in the field β_a, at which $2r_c(\beta_a) \approx a$, as shown by modeling, the middle part of the layer passes into the spin-flop phase by a first order phase transition. The field value β_a is [6]

$$\tau_a \equiv 1 - \frac{\beta a}{\beta_1} \approx \left(\frac{r_s}{a}\right)^2. \quad (8.54)$$

(b) $a \gg r_s \gg R$

This case corresponds to the area of weak vortex distortions. As shown in Sect. 8.4, in the field $\tilde{B}_R$ (formula (8.32)), weak distortions and a 90° exchange spiral occur near each layer surface. Then in the field β_a there is a spin-flop transition in the volume of the layer. This scenario is implemented at $\tau_a < \widetilde{\theta_R}$, which corresponds to the condition $aR > r_s^2$.

In the opposite case $aR < r_s^2$, there is a different course of events. In a particular field $\tilde{\beta}$, the entire volume of the layer immediately passes into the spin-flop phase with weak vortex distortions near the surfaces. The corresponding value $\tilde{\tau}$ can be estimated from the following considerations: the characteristic gain in power due to weak distortions is on the order of magnitude Rb_x / r_s^2 per cell of the layer surface. The loss in power (per cell of the layer surface) due to the occurrence of a spin-flop phase in a field less than the critical field β_1 is equal in the order of magnitude $\frac{a}{b_z}\left(\beta_1^2 - \beta^2\right) \approx ab_x\tilde{\tau} / r_s^2$. Therefore

$$\tilde{\tau} \equiv 1 - \frac{\tilde{\beta}}{\beta_1} \approx R / a. \tag{8.55}$$

(c) $R \gg a \gg r_s$

In this case of minor surface roughness, a scenario similar to the case $R \gg r_s$ for a thin layer is realized: as the magnetic field increases, a reorientation transition occurs sequentially in *OB*-type areas in the field β_{reor} (formula (8.36)), a spin-flop change in EA-type areas (EB) in the field β_s and spin-flop transition in the *OA*-type areas in the field β_1^*. The essential differences from the case of a thin layer are as follows:

(1) In the field β_s in areas of type *EA(EB)* in the center of the layer, there is a 180° domain wall parallel to its boundaries, which has a width of about r_s and occupies only part of the thickness of the layer. Recall that only the central part of the domain wall was placed in the thin layer. As the field increases and approaches the field of the bulk spin-flop transition, the wall captures the entire thickness of the layer due to divergence r_c. It happens in the field β_a. With further growth of the field, the width of the domain wall ceases to grow, remaining finite at $\beta \to \beta_1$.

To demonstrate this fact, we turn to Eq. (8.42). In the one-dimensional case, in the limit $r_c \gg a$, it takes the form $\theta''_{\varsigma\varsigma} = 0$, which indicates a linear dependence of the angle of rotation of the sublattices in the volume of the layer. In the center of the layer, the angle $\theta = \pi / 2$, to the surface of the layer, changes by the value χ so that the magnetization of the upper atomic plane is an acute angle with the direction of the applied magnetic field. The value χ is found from the boundary condition (8.45): $\theta'_\varsigma = \pm 2\beta \sin\theta$, which is converted to the form

$$\chi = \omega\cos\chi, \tag{8.56}$$

where $\omega = a\beta / b_z \approx a / \sqrt{2}r_s$. When $\omega \ll 1$, the value $\chi \approx \omega$, and when $\omega \gg 1$, there is the ratio $\chi \approx \frac{\pi}{2}\left(1 - \omega^{-1}\right)$.

The domain wall splits into two almost 90° exchange helices localized near the layer surfaces in the fields $\beta > \beta_1$.

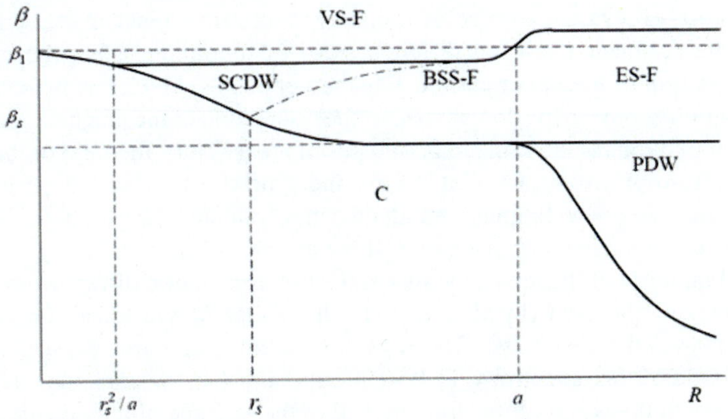

Fig. 8.15 "Magnetic field-roughness" phase diagram for a thick layer: BSS-F-phase with a surface spin-flop transition in *B*-type areas, SCDW-phase with surface 90° domain walls parallel to the layer; other designations are similar to those for Fig. 8.14

(2) A domain wall perpendicular to the surface of the layer and separating areas with an even and odd number of atomic planes, in fields close to β_1, has the following structure: its width δ near the surface of the layer containing the edge of the atomisc step is equal to $\delta_0 \sim b_x/\beta$. The width of the wall increases linearly as it moves away from the surface (Fig. 4.2), and $d\delta/dz \sim 1$ and reaches a value of $\delta \approx \min(r_c, a)$. In the case $r_c < a$, at further removal from the surface of the layer containing the step, the width of the domain wall does not change. In this case $r_c > a$, the growth of the wall width continues up to the opposite surface of the layer, and the power of such a domain wall is calculated based on the length of the unit cell on the surface of the layer of the order $\ln(a/\delta_0)$ [2].

In the fields $\beta \gg \beta_1^*$, the system's behavior is similar to that for the case of a thin layer. The phase diagram "magnetic field-roughness" of a thick layer of antiferromagnet is shown in Fig. 8.15.

8.6 Possibilities of Experimental Observation

Let us now focus on the possibility of experimental detection of the considered distortions of the antiferromagnetic order parameter. The position of the edges of the atomic steps on the surface of the antiferromagnet and the characteristic distance between them can be found by atomic force microscopy. Unfortunately, during the experiment, the parameter *R* usually remains in the shadow, and only the root-mean-square deviation of the surface from its average position is studied. Indirect estimates give a *R* value of the order of units–tens of nanometers. The improvement of technologies for the growth of crystals and nanolayers leads to a decrease in the concentration of atomic steps and growth *R*.

In the case of a rough surface of a massive antiferromagnet in weak magnetic fields in the collinear phase on different sides of the atomic step, the directions of magnetization of the upper atomic plane are opposite, which can be detected by magnetic microscopy. In the domain phase, the orientation of the magnetization of the upper layer of atoms is the same in both types of areas. Thus, the field of the surface spin-flop transition in the areas of the second type can be found experimentally. Magnetic microscopy techniques can identify the boundaries between the areas and make sure they coincide with the edges of the atomic steps.

At the boundary of the areas, the magnetization orientation differs from its value in the center of the area (in particular, in the center of the transition area, it is perpendicular to the easy axis). The width of the transition area δ_0, depending on the magnitude of the anisotropy, is 10–100 nm, and $\beta \leq \beta_1$ coincides in order of magnitude with the width of the domain wall in the absence of a magnetic field.

The phase with weak distortions can be easily identified by the reversal of the average magnetization of the surface layer of atoms from the direction of the magnetic field and the axis of easy magnetization. One can experimentally observe the transition from a phase with weak distortions to a vortex or domain phase in the fields $\beta > \beta_1$.

In the case of a nanolayer of antiferromagnet with uncompensated rough surfaces, the spin-flop transition leads to the appearance of some new types of domain walls, the widths of which depend significantly on the magnitude of the magnetic field. The observation of these walls by experimental methods is of undoubted interest from both fundamental research and possible practical applications.

Work on studying the surface spin-flop transition in the case of a naturally rough surface of an antiferromagnet is just beginning. And if specific steps have been taken toward the theoretical description of these phenomena, then experimental studies of the problem are still at the start.

References

1. Berzin AA, Morosov AI, Sigov AS (2005) Distortions of the magnetic structure induced by the magnetic field near the surface of the antiferromagnet. FTT 47(9):1651–1659
2. Berzin AA, Morosov AI, Sigov AS (2005) Dimensional effects in thin antiferromagnetic layers and multilayer magnetic structures of a ferromagnetic-nonmagnetic metal. FTT 47(11):2009–2014
3. Berzin AA, Morosov AI, Sigov AS (2006) Magnetic structure distortions induced by a magnetic field near the antiferromagnet surface. J Magn Magn Mater 300(1):153–158
4. Morosov AI, Morosov IA, Sigov AS (2006) Surface distortions of the magnetic structure of a uniaxial antiferromagnet: phase diagram “magnetic field-roughness”. FTT 48(10):1798–1804
5. Morosov AI, Morosov IA, Sigov AS (2007) Distortions of the magnetic structure of a thin layer of an antiferromagnet in a magnetic field. FTT 49(7):1228–1235
6. Morosov AI, Sigov AS (2010) Surface spin-flop transition in an antiferromagnet. UFN 180(7):709–722

Conclusion

Thus, we have shown that in the multilayer ferromagnet—antiferromagnet system, the behavior of the magnetic order parameters in layers of nanometer thickness is determined mainly by the frustrations that occur at the interfaces of the layers.

New types of domain walls generated by exchange interaction frustrations are predicted. The width of these walls is determined by the competition of exchange interactions within the layers and between the layers. And it turns out to be much less than the width of traditional domain walls.

The phase diagrams "layer (layers) thickness—roughness" of a thin ferromagnetic film on an antiferromagnetic substrate and a spin-valve system ferromagnet—antiferromagnet—ferromagnet are constructed taking into account the power of single—ion anisotropy.

It is shown that the presence of mutually perpendicular easy axes lying in the plane of the layers is favorable for the appearance of an exchange bias in the ferromagnetic—antiferromagnetic system. In addition, there must be a bound state of the domain wall at the interface, or there must be pinning of the domain wall by crystal lattice defects in the antiferromagnet near the interface.

It is a call on experimental physicists and technologists to determine the parameters of the interface as thoroughly and carefully as possible, and in particular, the parameter that is very important for the case of uncompensated atomic planes of an antiferromagnet parallel to the interface—the characteristic distance R between the edges of the atomic steps. Unfortunately, as a rule, this parameter remains in the shadow, and only the standard deviation of the interface from its average position is studied.

Only knowing the dependence of the value R on the technological parameters, it is possible to choose a technological route that will lead to the creation of a multilayer magnetic structure that corresponds to a particular area of one of the phase diagrams discussed above, which will allow one to achieve optimal parameters of magnetoelectronic components and devices. That, of course, requires in situ control of the specified parameter during the growth of the multilayer structure.

A. S. Sigov, *Multilayer Magnetic Nanostructures*, Springer Aerospace Technology,
https://doi.org/10.1007/978-981-19-6246-2